# Highways:
# An Architectural Approach

# Highways:
# An Architectural Approach

**Lester Abbey**

VAN NOSTRAND REINHOLD
New York

Copyright © 1992 by Van Nostrand Reinhold

Library of Congress Catalog Card Number
ISBN 0-442-00603-9

All rights reserved. No part of this work covered by the copyright hereon may be reproduced or used in any form by any means—graphics, electronic, or mechanical, including photocopying, recording, taping, or information storage and retrieval systems— without written permission of the publisher.

Printed in the United States of America.

Van Nostrand Reinhold
115 Fifth Avenue
New York, New York 10003

Chapman and Hall
2-6 Boundary Row
London, SE1 8HN, England

Thomas Nelson Australia
102 Dodds Street
South Melbourne 3205
Victoria, Australia

Nelson Canada
1120 Birchmount Road
Scarborough, Ontario M1K 5G4, Canada

16 15 14 13 12 11 10 9 8 7 6 5 4 3 2 1

**Library of Congress Cataloging-in-Publication Data**

Abbey, Lester.
    Highways: an architectural approach / Lester Abbey.
      p.   cm.
    Includes bibliographical references and index.
    ISBN 0-442-00603-9
    1. Roads—Design and construction.   2. Roadside improvement.
3. Landscape architecture.   I. Title.
TE175.A34   1992                                  92-1398
625.7—dc20                                           CIP

# Contents

**Acknowledgments** xi

**Preface** xiii

**Part 1** **Planning** 1

**Chapter 1** **Overview** 3
IS THERE REALLY A NEED FOR MORE HIGHWAYS? 4
COMMUNITY INFRASTRUCTURE 5
A LENGTHY TIME FRAME 6
THE PLANNING EFFORT 6
PRIVATE SECTOR INVOLVEMENT 10
PROGRAMMING 11

**Chapter 2** **Alternatives** 14
ARCHITECTS, PLANNERS, AND DESIGNERS 14
A PAUSE TO REVIEW 15
VALUE-ARCHITECTURE 17
COST-EFFECTIVENESS 20
ALTERNATIVES ANALYSIS AND REVIEW 21

**Chapter 3** **Multiple Use and Joint Development** 23
MULTIPLE USE OF HIGHWAY RIGHT-OF-WAY 23
JOINT DEVELOPMENT 29

vi    Contents

## Part 2    Preliminary Design    33

**Chapter 4**    Mapping    35
SURVEYING    36
PLOTTING CARTOGRAPHIC FEATURES    38
RECONNOITERING    46
FULFILLMENT    47

**Chapter 5**    Construction Materials and Foundations    48
SUBSURFACE    48
SOILS AND GEOTECHNICS    49
VALUE-ARCHITECTURE IN BASE CONSTRUCTION    55
SPECIAL AND UNUSUAL CASES    55
SOILS INVESTIGATION: FULFILLMENT    56
AGGREGATES    57
BITUMINOUS PRODUCTS    60
PORTLAND CEMENT    62
METALS    62
GEOSYNTHETICS    62

**Chapter 6**    Water    64
LOCATING THE HIGHWAY    64
WATER ENVIRONMENT    66
ECOSYSTEMS    67
WATERWAY DIVERSIONS    69
CLOSURE    73

**Chapter 7**    Pavements    75
CLASSIFICATIONS    76
FLEXIBLE    78
RIGID    80
MAKING THE CHOICE    81
MODULARS AND SYNTHETICS    82
TO SUMMARIZE    82
BENEATH THE SURFACE    83
FROST    84
DESIGN    85

REHABILITATION    86

**Chapter 8**  **Environment**  **91**
ENVIRONMENTAL ENHANCEMENT    91
"IT'S THE LAW!"    95
MITIGATION MEASURES    97
INTERDISCIPLINE OR MULTIDISIPLINE    101
ENVIRONMENTAL FULFILLMENT    101

**Chapter 9**  **Geometry**  **103**
PRELIMINARY LINE AND GRADE CONSIDERATIONS    103
PLOTTING THE ALIGNMENT    104
FINAL GEOMETRY    113

## Part 3    Final Design    123

**Chapter 10**  **Earthwork**  **125**
CLEARING THE WORK SITE    125
EXCAVATION    126
ESTIMATING QUANTITIES    128
SOIL ON THE WORKSITE    130

**Chapter 11**  **Drainage and Erosion Control**  **131**
MAJOR STEPS IN DRAINAGE DESIGN    132
EROSION CONTROL    137
DRAINAGE AND EROSION CONTROL: FULFILLMENT    141

**Chapter 12**  **Structures**  **144**
MAJOR TYPES OF MAJOR STRUCTURES    145
INNOVATIVE APPROACHES    149
ADDITIONAL STRUCTURAL ELEMENTS    149
REPLACING VS. RESTORING    150
PROBLEM AREAS    151
RETAINING WALLS AND SYSTEMS    152
INITIAL DATA REQUIRED BEFORE A STRUCTURE
  IS DESIGNED    154
GLOSSARY    156
IN CLOSING...    157

viii    Contents

Chapter 13   **Intersections and Interchanges   159**
    INTERSECTIONS   159
    HERESY   167
    INTERCHANGES   170

Chapter 14   **Roadside   176**
    VEGETATION   176
    WILDLIFE   180
    UTILITY RELOCATIONS   184
    ACCESS   186
    NOISE ABATEMENT   187
    REST AREAS   188
    INTERMODAL TRANSFER   192
    COMMUTER PARKING LOTS   195
    OTHER ROADSIDE MATTERS   197

Chapter 15   **Right-of-Way   200**
    ACCESS   200
    ROLE OF THE ARCHITECT   201

Chapter 16   **Plans, Specifications, and Estimates   203**
    PLANS   204
    SPECIFICATIONS   207
    ESTIMATES   209
    FINAL PACKAGE   211
    POSTMORTEMS AND REFLECTIONS   211

# Part 4   Construction   215

Chapter 17   **Contracts and Risk Management   217**
    CONSTRUCTION CONTRACTS   218
    RISK MANAGEMENT   220

Chapter 18   **Construction Items   221**
    SCHEDULING   221
    SITE PREPARATION ITEMS   222
    EXCAVATION   223
    DRAINAGE AND EROSION CONTROL   226

|  | PAVEMENTS 226 |
|---|---|
|  | STRUCTURES 229 |
|  | ROADSIDE 229 |

| Chapter 19 | **Safety** 232 |
|---|---|
|  | WORK ZONE SAFETY 232 |
|  | SAFETY AFTER THE FACILITY IS OPENED TO TRAFFIC 237 |

| Chapter 20 | **Inspection** 240 |
|---|---|
|  | STAFFING FOR A CONSTRUCTION PROJECT 240 |
|  | SITE PREPARATION 241 |
|  | EXCAVATION, BACKFILL, AND EMBANKMENT 242 |
|  | DRAINAGE AND EROSION 245 |
|  | PAVEMENTS 247 |
|  | STRUCTURES 251 |
|  | ROADSIDE 252 |
|  | THE GRAY AREAS OF INSPECTION 252 |

## Part 5   Operation   255

| Chapter 21 | **Traffic** 257 |
|---|---|
|  | ACCOMMODATING THE FLOW OF VEHICLES 257 |
|  | HUMAN FACTORS 260 |
|  | NONMOTORIZED TRAFFIC 262 |
|  | WEATHER CONDITIONS 263 |
|  | MANAGEMENT OF TRAFFIC OPERATIONS 264 |
|  | RAILROAD CROSSINGS 267 |

| Chapter 22 | **Maintenance** 271 |
|---|---|
|  | MAINTENANCE PRIORITIES 272 |
|  | PAVEMENT ACTIVITIES 274 |
|  | DRAINAGE 280 |
|  | BRIDGES AND MAJOR STRUCTURES 281 |
|  | ALONG THE ROADSIDE 283 |
|  | MAINTENANCE MANAGEMENT 286 |
|  | IN CONCLUSION 286 |

Chapter 23  Intelligent Vehicle/Highway Systems (IVHS)   289
            A CONCEPT COMING TO FRUITION   289

Index   293

# Acknowledgments

Although many people provided input for this book, the largest amount of guidance I received came from Jacques Chenet and William Marshall, civil engineers with the New York State Department of Public Works, where I began my career, and from Jonathan Palmer, a landscape architect with the Utah State Department of Highways, who introduced me to many of the concepts involved in highway architecture.

I am also indebted to Professor William Pollard of the University of Colorado at Denver and to Ed Patience, Division Manager DMJM, for reviewing the manuscript and offering suggestions. Also, many thanks to Robert Argentieri and his staff at VNR for the encouragement offered in pursuing this effort.

**PHOTOGRAPHS**

Where not otherwise indicated, photographs were taken by the author.

# Preface

What is highway architecture?
Who are the highway architects?
Where do they practice?
What is their role?

## WHAT IS HIGHWAY ARCHITECTURE?

Highway architecture is a way of attempting to achieve the best of both worlds by shepherding a highway project from planning through design, construction, and operation. It is an approach to rebuilding our highway infrastructure, from a humanistic rather than strictly an engineering point of view. Continuity of purpose is the prime objective. A corollary goal is to make the highway an integral part of its setting.

As now practiced, the building or rebuilding of any one highway is partitioned, fragmented, and compartmentalized. Planners hand a concept to designers; designers then prepare plans and specifications and pass their work on to construction people; construction people build the highway and turn it over to maintenance personnel. Rarely does one find continuity from planning to operation of a facility.

## WHO ARE THE HIGHWAY ARCHITECTS?

Although it is unlikely that anyone hands out a business card with occupation listed as "Highway Architect," this does not mean that no one practices the profession. Highway architects are those people who share the responsibility for developing a highway project. True, the practice is quite limited, but site development entrepreneurs, rural county engineers, landscape architects, and consultants to smaller local governments often perform as highway architects. They take a project from conception to completion and are concened with how the local community will react to it.

Some major state highway agencies are now using the project manager concept, whereby one person is responsible for guiding a project through both design and construction. This is not exactly highway architecture, but it's a start. One difficulty with such an approach is that the project manager is usually an engineer and, as such, will continually emphasize the technical side, sometimes at the expense of human values. If this same engineer were involved in the planning phases and in the environmental documentation, a considerably broader outlook could be engendered. Here is where the philosophical values implied by highway architecture can be of use.

Potential candidates for future highway architects could come from the ranks of:

- Civil engineers with a societal outlook;
- Professional and business people who have an interest in transportation;
- County highway superintendents;
- City road commissioners;
- Landscape architects;
- Site planners;
- Surveyors bent on expanding their horizons;
- University professors who are fed up with academia;
- Entrepreneurs who have become inured to bureaucracy; and
- Almost any reasonably well-educated creative person who is willing to embark on a new profession.

## WHERE DO HIGHWAY ARCHITECTS PRACTICE?

Highway architects practice in state and local highway agencies, in consulting firms, on construction projects, on road maintenance crews, on survey parties, in testing laboratories, and in other such sundry places that are frequented by members of the "highway community." Unfortunately, their exposure to highway architecture is very often limited to their own specific duties so that they never get a chance to view the whole picture—something like repeatedly studying the Mona Lisa's eyebrow or ear lobe through a magnifying glass.

## WHAT IS THE ROLE OF THE HIGHWAY ARCHITECT?

The role of the highway architect is very much akin to the role that a traditional architect (this term is used to distinguish them from landscape and preservation architects) fills when planning, building, and operating a major edifice.

A traditional architect will usually assemble a staff that is augmented by specialists. The staff may be comprised of a CADD operator, a detailer, a specifications researcher, an estimator, and several additional technicians suited to the task at hand. Specialists and consultants might include an acoustical expert, a site planner, plus structural and HVAC engineers. It is the same with the highway architect, who could employ a staff consisting of a CADD operator, a geometric-layout person, a hydraulics engineer, a utilities coordinator, and other needed technicians and inspectors. Outside aid could come from a mapping specialist, an environmentalist, a structural designer, a geologist, and a traffic engineer. Because the range of activities is so broad, a highway architect need not be a specialist, but should have a modest degree of knowledge concerning major facets of highway development.

Most traditional architects do not get a chance to design and build a Taj Mahal or a Hagia Sophia. Their bread and butter comes more often from designing and building warehouses, foundries, apartment complexes, slaughterhouses, civic arenas, and office buildings—functional but mundane.

Likewise, few highway architects get to design and build a Linn Cove Viaduct or a Glenwood Canyon Freeway. The meat and potatoes usually get paid for when the highway architect increases the capacity and enhances safety aspects of an existing highway, street, or intersection—again, functional but mundane.

Should either architect (traditional or highway) be able to improve the aesthetics and have the resulting edifice fit nicely into its surroundings, so much the better. In fact, here is where the architect faces the most interesting of challenges—the enjoyable part of the job.

## SOURCES OF INFORMATION

Obviously, a book of this size contains only a small amount of the information that is needed by a practicing highway architect. (References at the end of each chapter extend the depth a bit.) However, there are three prime agencies that continually provide up-to-date guidance materials for the practicing highway professional:

- American Association of State Highway and Transportation Officials
  444 North Capitol Street, NW
  Washington, D.C. 20001
  (202) 624-5800
- Federal Highway Administration
  Turner-Fairbank Highway Research Center
  6300 Georgetown Pike
  McLean, Virginia 22101
  (703) 285-2144
- Transportation Research Board
  National Research Council
  2101 Constitution Avenue NW
  Washington, D.C. 20418
  (800) 424-9818

All three agencies can be useful for networking.

Two centers that keep up-to-date information on highway-related software are:

- PC-TRANS, University of Kansas
  2011 Learned Hall
  Lawrence, Kansas 66045
- McTrans, University of Florida
  512 Weil Hall
  Gainesville, Florida 32611

*Site Engineering for Landscape Architects* by Steven Strom and Kurt Nathan is a good primer for someone who does not have a technical background.

## METRIC (SI) UNITS

In 1866 (that's 1866, not 1966), the U.S. Congress sanctioned use of the metric system, the only measurement system Congress ever sanctioned. Further legislation was passed in 1975 to encourage voluntary acceptance of metric units. With passage of the Omnibus Trade Act of 1988, mandatory adoption of the International Standard (SI) system was effected with a goal set for complete conversion by 1992. Why then are the units still spelled out in the English (British Imperial) system? Even the English no longer use the English system.

The response to this question has to be framed in the context of the nature of the highway community—a very conservative bunch. By the year 2000 (or maybe 2010), the highway people will be dragged (kicking and screaming) into the metric era. The Federal Highway Administration conversion is scheduled to be complete by 1996, but British Imperial units will likely be used beyond that date.

## TO GET ON WITH HIGHWAY ARCHITECTURE

Paul G. Edgecomb, chairman of the Oregon State Board of Landscape Architects, has made a most cogent observation in support of highway architecture:

> The highway system is perhaps the most influential cultural artifact of our time. Through its form, it seems to be a symbolic representation of our values. It might be wise to consider the aesthetics of this architecture. After all, we are what we design.

### References

Poister, Theodore H., Lloyd G. Nigro, and Randall Bush. 1990. *NCHRP Synthesis 163, Innovative Strategies to Upgrade Personnel in State Transportation Departments.* Washington, D.C.: Transportation Research Board, National Research Council.

Strom, Steven, and Kurt Nathan. 1985. *Site Engineering for Landscape Architects.* Westport, Connecticut: AVI Publishing Company, Inc.

# Part 1
## Planning

# 1
# Overview

Before getting bogged down in project details, the highway architect ought to look over—that's *look over,* not *overlook*—the full scope of the project concept. (Come to think of it, the highway architect should never get bogged down in project details.) One should view what might be termed the "big picture," before starting to fill in the specifics. In viewing the big picture, a logical place to begin is in the planning phase. As in most disciplines, a hierarchy subsists to the extent that there are three strata in transportation planning:

1. Policy;
2. Systems; and
3. Project.

The third item is of particular interest. Project planning is where the architect would be directly engaged. Just to confuse the issues further, we must also consider planning for the:

1. Long range;
2. Intermediate range; and
3. Short range.

The highway architect needn't worry about the first two. Short-range plans, having a horizon of five years or so, are the only ones that have any meaning. Moreover, they are expected to fit within the current socioeconomic fabric. Paradoxically, a highway rarely gets completed within its short-range planning time horizon.

## IS THERE REALLY A NEED FOR MORE HIGHWAYS?

Proponents of spending money for more and bigger freeways, expressways, highways, and roads continually point to urban congestion and rural accident rates to back up their cause. In rebuttal, their adversaries cite other (arguably better) ways to spend taxpayer and private money. This particular debate is one that a highway architect should avoid. In order to have a reasonably well-balanced, objective view regarding a project placed under his or her stewardship, the architect ought to stand on the sidelines and hold coats for the advocates of both viewpoints.

One aspect that highway proponents rarely proffer is the human need for a certain amount of privacy and seclusion. Many commuters find that the time spent going to and from work in an auto is the only time available to themselves. A typical suburban scenario may include an episodic family breakfast—baby crying, Junior procrastinating while preparing for school, spouse also readying for work, radio blaring news events. Then, during the workday, continual frustrations, turmoil, stress, and the normal traumas that come about when dealing with people. At days end, it's home again for an active evening with family, a rapid-paced TV sitcom, and maybe some sort of social or community activity. If the same commuter has to participate daily in a car pool with, for example, a cigar smoker or a chronic talker, plus the car radio tuned to assuage a fan of hard rock music, or if the commuter must travel by mass transit with its bumps and smells and periodic disturbances and hubbub, the commuter winds up with virtually no personal time—a good prescription for going bonkers.

There is an additional argument, rarely put forth on behalf of the continued building of more expansive urban-suburban traffic arteries: the multipurpose

**FIGURE 1-1.** Recommending solutions to urban traffic congestion seems to be governed most by the extent of one's involvement in urban traffic congestion.

commuter trips. These involve shopping, marketing, running errands, and depositing toddlers and elderly at day care centers and then retrieving them on the way home, a legitimate excuse for a one-occupant commuter vehicle. These subjective factors do not seem to be considered by those well-meaning souls who wish to get more folks into public transit, so as to remove single-occupant vehicles from suburban-urban freeways.

On the other hand, countering the standard arguments of the highway proponents, are the often miscalculated life-style changes of a population. In the future, will people continue to commute to and from work? Will their destinations always be to high-density centers? Will their working hours remain as they are now? Although predictions of things to come are necessary for any future planning, they rely too often on extrapolating past trends, thus, never quite fulfilling expectations. Is it possible, with the widespread availability of computers, modems, and fax machines, that home work schedules will obviate the daily round-trip downtown? Our life-styles are not static, so the one thing that the opponents of highways (more and bigger) have going for them turns out to be the unknown. Futurists contend that in the 21st century changes will occur in the work places, leading to changes in the nature of the work trip. Flextime and flex-workplace are concepts with which we must contend.

In order to advance their causes, both advocates and opponents of better highways have presented exaggerations and semi-truths. Some are briefly examined in the pamphlet *Myths and Facts about Transportation and Growth*, put out by the Urban Land Institute. A slightly different perspective can be found in *Transportation Research Record 1237*, which contains nine papers dealing with highway planning, attitudes of auto users, land use demographics, and related topics. The papers display the variety of approaches, conflicts, and dilemmas being explored by those who are attempting to help us plan our way out of existing and projected transportation shortcomings.

The highway architect is advised to have at least a smattering of interest in the direction that transportation planning is going, but not to the extent of getting so emotionally involved on either side of the many issues that one's sound and rational judgment is impaired. In this treatise, the lead hypotheses will be that most existing highways will, at some time in their life cycle, need modifications and rebuilding and that some new highways will be designed and constructed. The *Intermodal Surface Transportation Efficiency Act of 1991* (ISTEA) supports these hypotheses.

## COMMUNITY INFRASTRUCTURE

There is a human tendency to ignore those things that do not stand out. The brightly colored cereal package is going to sell better than the plain-packaged brand. Most of a community's infrastructure is not readily visible: sewers, buried cables, water aqueducts, gas mains, and the like. Civic leaders and political figures recognize that what cannot be seen will rarely be a major public concern. Such a viewpoint holds true until a clogged sanitary sewer habitually backs up, or a buried telephone cable is cut, or a culinary water aqueduct becomes contaminated, or a high-pressure gas main bursts. These are the times that public concern also bursts, yet no one is accountable.

Highways are part of a community's infrastructure, but are much more visible; therefore, they turn out to be a regular major public concern. When traffic habitually backs up, when a route is cut off, when an airshed becomes contaminated with carbon monoxide, or when a pavement slab bursts, public concern manifests.

Sewers, buried cables, water aqueducts, gas mains, and highways are all part of the urban and suburban infrastructure. In rural settings, the road network and power lines are the principal components of the infrastructure.

The intrusive factor with virtually all infrastructures is that they rarely receive adequate maintenance. "If it ain't broke, it don't need fixin." Since such philosophy is pervasive, political leaders and other decision makers tend to wait until a system breaks down completely rather than take any measures to prevent the breakdown. After all, they rationalize, the public will question, closely scrutinize, and sometimes protest an expenditure of $100,000 to keep an infrastructure component working, but that same public will approve without question $10,000,000 to fix that component when it collapses. It is here that the highway architect becomes involved in a different jousting tournament than that of the traditional architect or landscape architect. The playing field isn't quite so level.

## A LENGTHY TIME FRAME

It takes about one decade for a major highway construction or reconstruction to gestate (i.e., from initial planning to opening to traffic). Several more years are involved in establishing operational and maintenance strategies (e.g., modifications to signing, correcting localized slope erosion, checking traffic signal warrants). For any single individual to remain involved with a project for 12 to 15 years is not the norm. Yet there should be an orderly transition when a project shepherd is replaced, preferably *during* a phase rather than between phases. It is better for a replacement highway architect to take over in the middle of the design effort, for instance, rather than just before construction begins. This will better allow most of the project history to be overlapped.

## THE PLANNING EFFORT

A broader sweep of attention has been working its way into the planning process. Whereas, in the past, planners' concentration centered on the capacity of highways, networks, and corridors, the new thrust also embraces:

- Advanced technologies;
- Multimodal transfers;
- Environmental concerns;
- Energy conservation;
- Incident management;
- Access control;
- Corridor conservation;
- Cost effectiveness; and
- Demand management.

Each region of the country has its own relative priorities for these factors, but the data and other information collected for and placed at the disposal of the planners continues to be mainly of a historical nature. Traffic and its ancillaries still dominate the thinking because vehicle counts, sign inventories, skid measurements, accident analyses, bridge adequacy ratings, pavement conditions, and geometric shortcomings comprise the greater part of the statistical material that planners have to work with.

**Terms and Jargon**

Although architects are often involved in the planning stages, it isn't until the preliminary design phase that a highway actually gets "architectured." Nevertheless, the highway architect needs to know some of the catch phrases, vocabulary, and acronyms rampant throughout the transportation planning process.

The initiatory buzzword term is *average daily traffic* (ADT). This is simple enough in itself: the number of vehicles a highway carries in a 24-hour period. But the planners like to break it down further. The term *average annual daily traffic* (AADT) is used when seasonal variations are apt to skew the numbers, such as traffic near resort areas. More common is *average weekday traffic* (AWDT), which is limited to Mondays through Fridays. Urban locales where commuter traffic dominates are most subject to the AWDTs.

Hourly traffic counts are often treated with disdain by planners. They like to use a "$k$ factor," applied to the ADT (or the aforementioned siblings of ADT). A $k$ factor is usually somewhere close to 0.10 in urban and suburban areas. This means that the highest hourly traffic during the day is 10 percent of the total daily traffic. A $k$ factor rarely goes below 0.07, even on urban streets that are busy all day long (except in southern California where rush hour traffic goes on all day and half the night, resulting in a $k$ factor near 0.05). However, on lightly traveled rural roads, it is not unusual to come across a $k$ factor as high as 0.25; one-quarter of the total traffic in a day occurs during a one hour period.

*Directional split* is important, particularly on suburban arterials. A 60-40 split means that, at any one time, 60 percent of the traffic is going one direction, while 40 percent is trying to get to where the 60 percent came from.

*Design hourly volume* (DHV) was once the common approach to defining maximum traffic that one could expect on a highway segment. It isn't really the maximum, but rather the 30th highest hour of the year. (Most years have 365 × 24 = 8760 hours, so the DHV is pretty close to the maximum). *Peak hour volume* has supplanted DHV as a planning device. More subjective and site-specific, peak hour volumes are usually derived graphically or by computer program, where the higher volumes tend to level off, perhaps at the 200th highest hour of the year.

### Rural Issues

Even though the largest expenditure on highways takes place in urban and suburban locales, the rural segment, on a per capita basis, is at least as significant. The interstate component of the national highway system is the backbone of the rural transportation network and farm-to-market roads are also important, but local roads are deteriorating much faster than they can be revivified. In addition, the aging rural population will need better public transit. And to warm the heart of the architect, scenic byways are gaining favor.

### Suburban Issues

One reason people move to the suburbs is to get away from high-density living. Because travel distances from home to market and shopping locations often extend beyond comfortable walking limits, the automobile is the preferred mode of travel. With so much of the suburban lifestyle structured around the auto free parking is considered a basic right guaranteed by the Constitution. Inordinate emphasis

**FIGURE 1-2.** Interstate highways have become the fundamental feature of rural transportation.

toward accommodating motorized travel has had a significant negative impact on land uses and on other modes of getting around in the suburbs. Very few site designs accommodate pedestrian travel, outside of shopping malls and office complexes. In fact, some parking lots are extremely treacherous for those who embark from their vehicles to walk the short distance to the mall.

A highway architect might elect to go along with the tide, tacitly accepting suburbia for what it has become, but the greater satisfaction comes from attacking the challenge—not from the planning perspective, but from the design viewpoint. After all, something ought to be done to salvage suburbia. Why not begin with the highway architecture?

## Urban Issues

Too much has been written and spoken about urban transportation problems and alleged solutions. Not much has been accomplished in solving urban transportation problems. It was once thought that, when congestion is the problem, capacity is the solution. Not correct. It seems that, every time capacity is increased, congestion becomes more of a problem. Every time capacity is reduced, congestion becomes more of a problem. And every time capacity remains unchanged, congestion becomes more of a problem.

Combining the carrot with the stick is a popular approach. Some municipalities are offering incentives (carrots), such as transportation credits in the paycheck for those using public transit regularly instead of their auto. Free fuel for vanpools and subsidized bicycle and footwear maintenance are being tried. And then there is the other side of the pizza (the stick). Restricting everyday activities is being attempted in a number of heavily congested urban areas. Banning trucks, limiting the number of cars that a family may own, tripling registration fees for owners of a second car, and charging more for parking if the vehicle has only one occupant are some of the other means that are being attempted, with very limited success so far. Not to get discouraged, several approaches have shown a glimmer of success—or at least a potential for success—in alleviating some of the urban transportation hang-ups.

## Special Lanes

Traffic lanes reserved for buses and high occupancy vehicles (HOV) have been adopted in more than two dozen urbanized areas as part of their transportation management programs. The verdict is not yet in on how universal their applicability would be. They do show promise, though, and may increase in favor as the public gets more fed up with gridlock.

During rush hour, several communities are attempting to use freeway shoulders (10 feet in width) as additional lanes. The difficulty, of course, is when the on and off ramps are crossed. However, using the shoulders for small vehicles only, including electric cars, may have real potential for the future.

**Park and Ride**

One compromise to assuage both those who like a few moments of solitude in the privacy of their own automobiles and those who want to reduce traffic congestion is the installation of commuter parking lots. These are not exactly a new phenomenon, but historically they have been most successful where a change in travel mode is involved. Parking lots at commuter rail stops are the most easily recognized example.

A more recent application of commuter parking lots involves no change in travel mode: auto to auto (or van or bus)—that is, a car pool rendezvous. The philosophy is to provide strategically placed small-scale, low-cost facilities that obviate the need for extra lanes on existing arterials leading to high-density employment centers, whether they be downtown or suburban. The key is to make these commuter parking lots attractive enough so that they will achieve high use. Without high use, they would be just a waste of capital that could better be expended elsewhere. Later on we'll go into some of the measures that can be used to make commuter parking lots an effective tool in combating impending gridlock.

A complementary action is to make use of existing parking lots as a surrogate for building new ones. It often benefits all concerned. A study done in Montgomery County, Maryland, (Smith, 1983) summarized:

> ...there can be a significant economic benefit to shopping-center operators for allowing commuter parking to occur on their parking lot. Survey results indicate that between 25 and 45 percent of park-and-riders shop at the shopping center on a typical day on their way to or from work. Approximately two-thirds of this shopping activity is either diverted from other shopping locations or is newly induced shopping. For the shopping centers surveyed, the average increase in sales due to the presence of park-and-ride activity is $5/park-and-rider/day. Also, the presence of the park-and-ride facility, in itself, is responsible for 10–30 percent of the park-and-riders choosing transit or carpooling.

This shows that the shopping center is an effective tool for attracting park-and-riders. The charge to the highway architect, presuming that use of an existing parking lot has been granted, is to provide convenient access and egress.

## PRIVATE SECTOR INVOLVEMENT

Throughout the twentieth century it has been the governmental entities that have provided the major highways. As the next century approaches, private capital will begin to play a more significant role. Private capital has been a main source for local streets where residential subdividers and industrial developers are required to build the roads and utility installations within the subdivisions or developments and then turn the streets over to the municipal governing body.

Privatization of our highway and freeway networks is already taking place. The E-470 Roadway is actually a beltway (well, half a belt) along the plains east of Denver. Four local governmental units and a group of landowners and developers are financing this major toll road. Projected shortfalls of revenue during the first few years of operation are to be offset by some imaginative financial arrangements. As more private entrepreneurs get into highway development, we may expect many unique financing methods to become commonplace. In the meantime, the highway architect must deal with the more conventional approaches to pay for highway improvements.

## PROGRAMMING

Assigning funds to highway projects and setting priorities does not usually fall under the responsibilities of planners; the matter is too sensitive. Planners, as a courtesy, are allowed to provide input, but spending public or private monies is within the province of political and entrepreneurial masters of the game. The planning process may have been heavily loaded with professional technological analyses and factual data, yet the actuation of these plans, through programming, becomes more of an art form. It is best for the highway architect to realize this truism early on so that any questions as to why a project was conceived and programmed in a certain way can be placed in a proper frame of reference.

### Goals of Society

At any given moment, a social order has certain objectives. Highway concepts, as well as their financing, have not escaped societal goals. Some examples of using highways to accomplish societal goals have included:
- Managing growth;
- Attracting employment generators for minority groups;
- Controlling land use;
- Redirecting environmental issues;
- Using the Highway Trust Fund to reduce the federal deficit;
- Accommodating localized economic expansion;
- Restricting localized economic expansion;
- Increasing wetland acreage;
- Preserving historic sites; and
- Obliterating old neighborhoods.

Consequently, one advantage of having the highway architect participate in the planning phase of a highway project is to obtain an awareness of which immediate socioeconomic objectives are to be included in the project. Decisions made during project development can then be made within the proper framework.

## Data Envelopment Analysis

Title I of the *Intermodal Surface Transportation Efficiency Act of 1991* (ISTEA) authorizes $121 billion for highways and related transportation. However, program funds for federal aid projects are now distributed in a different manner than in the past, requiring transportation agencies to revise some of their programming processes.

Programming personnel in several major transportation agencies are focusing their attention toward a concept derived from a technique put forth by academic management scientists. Data envelopment analysis, developed in the latter part of the 1970s for use in the public sector, assesses the relative "efficiency" of various projects under consideration. Evaluation of resources required to produce outcomes of proposed projects are compared so that each project can be rated as to its efficiency in accomplishing the goals of the transportation agency (Walters, 1992). Approaches to programming funds, such as this, are becoming necessary to allow decision making the opportunity to choose the best mix of budgeting and promoting societal objectives.

## References

Larson, Thomas D., and Kant Rao. 1989. *NCHRP Synthesis 151, Process for Recapitalizing Highway Transportation Systems.* Washington, D.C.: Transportation Research Board, National Research Council.

*Myths and Facts about Transportation and Growth.* 1989. Washington, D.C.: The Urban Land Institute.

Smith, Steven A. 1983. Park-and-ride at shopping centers: a quantification of modal shift and economic impacts. *Transportation Research Record 908*:27–31.

Transportation Research Circular 359. 1990. *Traffic Congestion and Suburban Activity Centers.* Washington, D.C.: Transportation Research Board, National Research Council.

Transportation Research Record 1197. 1988. *Transportation Finance and Economic Analysis Issues.* Washington, D.C.: Transportation Research Board, National Research Council.

Transportation Research Record 1236. 1989. *Geographic Information Systems: An Important Technology for Transportation Planning and Operations.* Washington, D.C.: Transportation Research Board, National Research Council.

Transportation Research Record 1237. 1989. *Congestion, Land Use, Growth Management, and Transportation Planning.* Washington, D.C.: Transportation Research Board, National Research Council.

Transportation Research Record 1243. 1989. *Future of Statewide Transportation Planning.* Washington, D.C.: Transportation Research Board, National Research Council.

Transportation Research Record 1285. 1990. *Transportation Forecasting.* Washington, D.C.: Transportation Research Board, National Research Council.

Walters, Lawrence C., and Glen S. Thurgood. 1992. *Prioritization Formula for Funding of Projects.* Provo, Utah: Brigham Young University, Institute of Public Management.

Wuestefeld, Norman H. 1991. Toll roads—private sector funding. *TR News* 155:2-5.

# 2
# Alternatives

A core element of all transportation planning is the review of alternatives to any single project. Quite often, a proposed action (which appears at first to be the best solution) gets mired in so many difficulties that some other action must be undertaken. When a project gets tied up in environmental or social issues after considerable design work has been done, hindsight says that a more thorough planning effort, which includes uncovering all the available options, could have obviated a trip "back to the CADD terminal."

## ARCHITECTS, PLANNERS, AND DESIGNERS

Planners, to their credit, tend to see the entire forest, but, in so doing, may miss counting some of the trees. The highway architect must see both the forest and all the trees. Having such a vantage, the architect is able to ensure that a highway concept can be carried to fruition without destroying the concept in the process. Not always so with designers. Many times a trusting highway designer will take a project concept that has been properly programmed by a sponsoring agency and, because the design schedule is tight, run with it like a frightened rhinoceros. Here is where the highway architect can provide direction and guidance to the designer, should the architect not personally take on the role of designer.

Before getting too carried away, so that the highway location and design get hopelessly locked in, it is good practice for a designer to consider any feasible alternatives to the programmed proposal. (One may ponder whether Captain Smith of the *Titanic* would have completed his mission had he considered an alternative routing—or an alternative to sailing full speed through the ice field.) Sometimes alternative actions have already been adequately analyzed in an associated environmental assessment or impact statement. There are situations where, in locating and designing a highway, obstacles (not necessarily icebergs) crop up unexpectedly. They may be difficult to overcome; therefore, it's prudent to have at least one fallback option so that the project does not get unduly bogged down or possibly scuttled.

From Point A to Point B, there are usually a myriad of possibilities. Notwithstanding an elementary logic, the shortest distance for a highway is not always a straight line. Rarely, in fact, is the straight line the most desirable routing—as shall be pointed out later on.

## A PAUSE TO REVIEW

Alternatives can be classed into five groups:

- Maintaining the status quo;
- Alternative types of facility;
- Alternative mode(s) of transportation;
- Alternative highway alignments within a corridor; and
- Alternative highway corridors.

In the larger highway agency bureaucracies, a designer is not encouraged to mingle with the planners. "They do their job; you do yours." Unfortunately, the designer is then not privy to some of the reasoning behind the programmed concept that he or she must cope with. At this point, the architect could step in and conduct an inquiry into just how far and to what extent the planners considered the array of available alternatives before proceeding further. NCHRP *Report 96* presents a lengthy overview of the transportation planning process. It can be used for reference should the architect need to take issue with how a selection of one or more alternatives was made by the planners. (Page 78 of the report shows how an electrician might diagram that which takes place in the planning process.)

Few circumstances require that the architect go back to the planners for justification of their methods. Yet the architect may wish to mull over whether any other options to the proposed action are viable in light of unforeseen problems that may arise during design of the project.

### Maintaining the Status Quo

There are several levels of the "No Action" alternative. In cases where an existing route is involved, they range from the generally unacceptable "Do Nothing," through the more pragmatic continuance of regular maintenance functions, to minor enhancements by incorporating a greater maintenance effort and several

16    Highways: An Architectural Approach

**FIGURE 2-1.** The "No Improvement" alternative requires an increasingly greater maintenance effort.

operational improvements, on a small scale, to prevent deterioration of the existing highway.

*Transportation systems management* (TSM) is one aspect of the "No Action" alternative that came into favor during the 1970s. A treatise on TSM can be found in NCHRP *Report 263*; a quick introduction is contained in NCHRP *Report 283*. (On page 6 of the latter publication is a list of many TSM "Action Profiles," with all except nos. 7 and 20 fitting into the "No Action" alternative, insofar as the highway designer is concerned.)

### Alternative Type of Facility

If applicable, could a different type of highway (two-lane highway, with an added high-occupancy vehicle lane (HOV), instead of a four-lane freeway; or limited access expressway instead of non-access collector road) do the job? If not, then why not? TR *Record 1081*, pages 8–82, contains a number of articles relevant to expanding road service capabilities.

At times, changes in driving patterns that affect the number of occupants per auto may significantly alter the capacity demands of a proposed highway. Also, a small expenditure on a specific ancillary feature (such as the installation of suburban commuter parking lots from which carpools can be conveniently expedited) may be able to reduce demand on the capacity of a highway enough to convert some projected major improvements into minor ones.

### Alternative Mode(s) of Transportation

Could air, rail, or water vehicles provide a commensurate level of service or be used to supplement vehicular traffic? Likewise, are other means of land transit practical,

**FIGURE 2-2.** Public transit is favored by many planners as an alternative to one-occupant commuter vehicles.

feasible, or available to replace, complement, or supplement the existing highway or portion of the proposed highway? Various possibilities involved with mass transit (particularly in urbanized areas) may be suitable as a surrogate for major road improvements.

### Alternative Highway Alignments Within a Corridor

A corridor is usually defined by topographical or developmental constraints that limit (or bound) the physical placing of an alignment. However, there might still remain many possibilities within any given corridor.

### Alternative Highway Corridor

Sometimes the frustration of attempting to overcome all the hurdles within a transportation corridor becomes excessive or insurmountable. Then it is time to ask: What are the available parallel corridors that could fulfill the transportation needs or function of the proposed highway? The architect can get in position to consider reasons why their use would be less (or more) satisfactory than the selected corridor and proceed from there.

## VALUE-ARCHITECTURE

Where public as well as private money is to be spent, the taxpayers or entrepreneurs expect a useful return. Value-Architecture—much like Value Engineering—has a point of view that asks: How much will it be worth?

Worth, in a technological sense, would be the least expensive manner in which to implement a function, giving minuscule consideration to just how the function is applied. Value-Architecture (Val-Arch) begins once most functions have been identified and have been classified, either as basic or secondary. By this time it may be assumed that nonessential functions have been discarded. Judgment wields a prominent role in Val-Arch, which is predicated on the proposition that money is spent to accomplish functions rather than simply to obtain ownership. It is at least an extension of good practice, better than a suggestion program, and goes beyond customary project or plan reviews. Since its aim is to maintain a satisfactory level of product or service, Val-Arch is rarely used to cut corners.

## What Else Could Accomplish the Goal of a Transportation Service?

Alternative considerations are usually required on major highway projects, but not necessarily at each stage in the normal day-to-day development of those projects. Val-Arch quite often requires a certain amount of expense that ought to be justified by potential cost savings. It follows that, before entering into a major Val-Arch alternative endeavor, there should be a recognized need or opportunity for financial benefit to the sponsoring agency.

Contemporary concerns for environment, energy, and rising costs lead to the conclusion that functional aspects related to safe and efficient transportation may have to be analyzed individually. In this way, important functions may be attained in an optimum economical setting. Value Engineering has been applied to the highway field contemporaneously with the drying up of financing sources. Consequently, throughout this treatise, essential points of Value Engineering as well as Value-Architecture will continually surface.

## When Does Val-Arch Get Involved in Highway Location?

Maximum effectiveness comes about when Val-Arch is undertaken early enough to affect and effect decisions on life-cycle costs. When basic criteria are being established, time taken to identify necessary and essential functions, to separate or drop unnecessary and unessential functions, to assemble an array of logical alternatives, and to pursue those cost-effective alternatives is time well spent. For instance, location considerations ought to make optimum use of existing value. Using portions of an existing facility is too often ignored when a designer is charged with upgrading or improving a highway. Should terrain and structural stability of an existing alignment be satisfactory, appropriate components of the old facility could well be incorporated into the upgrading. Some aspects that may readily lend themselves to this philosophy include:

- Right-of-way—Often inflated real estate costs place this facet as one that presents opportunities for obtaining value via the use of existing rights-of-way. Any improvement made to an existing roadway in a developed area should acquire minimal new right-of-way.
- Fills—Use of previously compacted embankments often leads to considerable savings in earthwork.
- Bridges—Instead of removing an old structure and replacing it with a wider one, savings might be had by retaining the existing bridge for one-way traffic and placing a modest parallel structure for the other direction of traffic.
- Recycling—Many times, existing pavements and roadbeds can be recycled and incorporated into bases and sub-bases for the new highway.

Another point, often overlooked, is the rigid application of excessively high standards to a facility that may not warrant too great of an expenditure of resources. (In NCHRP *Report 63*, Appendix A, a rationale is described for lowering design standards on low-travel rural roads. More ammunition along this line may be found in TR *Record 875*, pages 53-60.) Much of the available design-oriented computer software is programmed with the highest order of design standards. It would then seem that some judgments ought to be made by the designer when applying CADD (or other computer-generated) programs, rather than accepting their printouts as unalterable, especially when aspects of overdesign are evident.

During preliminary design, attention to value considerations would most likely be directed towards earthwork, pavement section, drainage, and structures. Other items are usually scrutinized further along at the appropriate step in the design activities. (NCHRP *Synthesis 78*, pages 8-12, presents a brief overview of how

**FIGURE 2-3.** Widening an existing highway: where an older bridge still has a serviceable life, it can be retained for traffic in one direction when a new structure is placed next to it for the other direction of travel.

Value Engineering has been applied in the highway agencies of California, Florida, Idaho, Minnesota, New Mexico, Oregon, Pennsylvania, and Virginia and by the U.S. Army Corps of Engineers.)

Alternatives (not necessarily limited just to alternative alignments) such as a choice of number of lanes, the option of going down one side or the other of a valley, or the variables involved in provision of interchanges and access, when considered early on in project development, are other Val-Arch approaches that can often lead to important savings of the taxpayer dollar.

## COST-EFFECTIVENESS

Once upon a time it was dogma to develop a benefit-to-cost ratio before pursuing the design of a major highway project. As available capital started drying up, it was seen that an initial investment large enough to obtain a desirable benefit-to-cost ratio (something well in excess of one) was often out of the question. Thus came the thrust toward cost-effectiveness, whereby a proposed transportation facility was required to justify the expense (within available funding), rather than to provide the highest benefit-to-cost ratio.

Just how far should a highway architect go in developing cost-effectiveness? How detailed should it be?

Few architects have the urge to become dedicated card-carrying economists. Yet a professional approach to cost-effectiveness is expected. As in Value-Architecture, where the greater emphasis is placed on items that would be expected to establish the greater value, while lower priorities are assigned to lesser value items, the architect should concentrate on costs that play a major part of the total picture. As it becomes necessary to refine the cost-effective analysis, smaller expenditure differentials are sought.

One way to begin is to set up a rudimentary cost matrix, as in Table 2-1.

In order to make fair comparisons, "present worth" must be considered and life-cycle spans established for the project alternatives. (Although a highway might be given an overall "life" of 25 years, its different components vary considerably. Pavement life may be only 15 years, while properly maintained structures are usually expected to last 50 years. Drainage schemes and pipes vary from 20 to 80 years, and the life of the transportation corridor itself may be more than a century.) Once all the appropriate blocks have been filled in, total annual cost for each alternative can be compared with "No Action" and with each other.

If an alternative turns out to be less expensive than the "No Action" option, it can be considered cost-effective. Also, a range of alternatives, some more expensive in initial cost than others, can be put in the matrix so that the decision makers can view how best to spend the available capital.

Here is where the architect can decide either to go back to a text in basic

**TABLE 2-1  Rudimentary Cost Matrix**

| Expenditure Phase | No Action | Alt. #1 | Alt. #2 | ... |
|---|---|---|---|---|
| *First Costs* | | | | |
| 1  Planning | – | X | X | |
| 2  Design | – | X | X | |
| 3  Construction | – | X | X | |
| *Continuing Agency Costs* | | | | |
| 4  Maintenance | X | X | X | |
| 5  Operation | X | X | X | |
| *User Costs* | | | | |
| 6  Time Value | X | X | X | |
| 7  Vehicle | X | X | X | |
| 8  Running | X | X | X | |
| *Assorted Costs* | | | | |
| 9  Environmental | – | X | (–) | |
| 10  Adjacent Property Values | (X) | X | | |
| 11  *** | | | | |

economics or to bring in a sympathetic economist to help prepare the cost-effective analysis.

## ALTERNATIVES ANALYSIS AND REVIEW

Upon completing a review of alternatives, the designer should possess at least one fallback option to accompany the project through much of the design phase. There is even the ecstatic possibility that all will go well and no potentially feasible alternative will have to be pursued at all.

However, before we begin to lay out any new or revised highway alignment, it is best to view the selected corridor (and alternative, if more than one corridor may still be under consideration) for optimum use of the available land. This leads us to the next chapter.

### References

Anderson, Dudley G., et al. 1977. *A Manual on User Benefit Analysis of Highway and Bus-Transit Improvements.* Washington, D.C.: American Association of State Highway and Transportation Officials.

Batchelder, J. H., et al. 1983. *NCHRP Report 263. Simplified Procedures for Evaluating Low-Cost TSM Projects.* Washington, D.C.: Transportation Research Board, National Research Council.

Campbell, Bruce, and Thomas F. Humphrey. 1988. *NCHRP Synthesis 142, Methods of Cost-effectiveness Analysis for Highway Projects.* Washington, D.C.: Transportation Research Board, National Research Council.

Mason, J. M. Jr., *et al.* 1986. *NCHRP Report 283. Training Aid for Applying NCHRP Report 263.* Washington, D.C.: Transportation Research Board, National Research Council.

NCHRP Report 96. 1970. *Strategies for the Evaluation of Alternative Transportation Plans.* Washington, D.C.: Transportation Research Board, National Research Council.

Oglesby, C. H., and M. J. Altenhofen. 1969. *NCHRP Report 63. Economics of Design Standards for Low-Volume Rural Roads.* Washington, D.C.: Transportation Research Board, National Research Council.

Technical Advisory T7570.1, *Motor Vehicle Accident Costs.* 1988. Washington, D.C.: Federal Highway Administration.

Transportation Research Record 875. 1982. *Design and Upgrading of Surfacing and Other Aspects of Low-Volume Roads.* Washington, D.C.: Transportation Research Board, National Research Council.

Transportation Research Record 1081. 1986. *Urban Traffic Management.* Washington, D.C.: Transportation Research Board, National Research Council.

Transportation Research Record 1280. 1990. *Transportation Management, HOV Systems, and Geometric Design and Effects.* Washington, D.C.: Transportation Research Board, National Research Council.

Turner, O. D., and Robert T. Reark. 1981. *NCHRP Synthesis 78, Value Engineering in Preconstruction and Construction.* Washington, D.C.: Transportation Research Board, National Research Council.

# 3
# Multiple Use and Joint Development

Looking down from an airplane window upon approaching any major airport, one can get an excellent perspective of just how much territory is taken up by interstate freeways and their attendant interchanges. Major arteries and local streets also occupy plenty of land. A fifth of the urban landscape, plus almost as much of the suburban landscape, is land dedicated to highway right-of-way.

One of the more valuable contributions an architect can make in highway project development is to search out and implement various other compatible uses for the large chunks of real estate that highways occupy. Surprisingly, planners, as well as engineers and designers, tend to disregard using highway corridors or portions thereof for anything but autos and trucks (and occasionally buses).

Although multiple use concepts, combined with joint development, have been around for many years, the fragmented nature of highway project development has prevented, in many regions and locales, any concerted effort to carry out the concepts as an overall policy matter. The decision makers try to avoid any additional nuisances that might interfere with their charge to keep the traffic moving. By default, therefore, the highway architect can and should assume the initiative and coordination requisite to see that appropriate and viable multiple use concepts are included.

## MULTIPLE USE OF HIGHWAY RIGHT-OF-WAY

The U.S. Forest Service has been a proponent of multiple use for a considerable period of time. They have permitted and sometimes encouraged compatible use of specific areas in national forest lands, where practicable (forest fires notwithstand-

**FIGURE 3-1.** Optimal use of the highway right-of-way in Duluth: the downtown freeway was enclosed so that the space above can now be used for local activities.

ing). Recreation, timber harvesting, watershed preservation, and wildlife habitat, when properly managed, can occupy the same land area to the benefit of many interests.

Too often, in the past, a number of highway agencies, in pursuing their stewardship responsibilities, forbade all other activities—compatible as well as noncompatible—from occupying the highway corridor or any part thereof. Times change and so do highway agencies and their policies.

Multiple use of highway right-of-way (above and beyond the primary use by the highway itself) utilizes highway right-of-way to centralize services in order to fully use available land. Centralization of services may aid and assist not only the travelers, but also the communities in which this concept is instituted. Of prime importance in multiple use is each community's land use objectives.

Focus of multiple use efforts was once directed to the economic definition of the "highest and best" use of land. Lately the philosophy has been shifting to other values, such as historic preservation, conservation of natural resources, and even aesthetics. Current desires and needs of the communities nearest the proposed facility are paramount; therefore, all multiple use of highway right-of-way proposals should be approved by the local governments and be in compliance with regional land use plans.

Because of the fact that comprehensive planning for land use, multiple use, and transportation are interdependent, planning efforts may have to be coordinated in the beginning, as well as throughout the highway project development process. Some difficulties may arise when coordinating with other agencies whose lead time for project development differs greatly from the up-to-ten years of gestation for highway planning, location, design, and construction. Also, multiple use efforts are best completed prior to the opening of the highway so as to minimize traveler inconveniences.

Above all is this important point: for all multiple use projects, the land used must in no way infringe on the safety of the highway facility or on routine maintenance operations.

## Opportunities for Multiple Use of Highway Right-of-Way

Few of the following schemes would apply to any individual project. Some would be applicable only if there are one or more larger parcels in a right-of-way taking. If those larger parcels would be highly impacted (rendering the remnant portions unusable by the owners) so that the highway agency purchases the entire parcel, the applicability of multiple use is enhanced.

Opportunities for agricultural utilization could include:

- Forestry (*Transportation Research Record 913* presents a case for right-of-way forestry on pages 14-18);
- Orchards;
- Grazing lands (properly fenced);
- Agricultural products (alfalfa); and
- Garden plots for the immediate neighborhood.

Many large barren interchange areas—particularly cloverleafs and directionals—can be retrofitted with emergency facilities such as:

- Highway patrol/police stations;
- Ambulance garages;
- Fire houses; and
- Heliports.

Sometimes, highway agencies overlook opportunities to spot their own facilities within existing or proposed right-of-way:

- Maintenance stations;
- Stockpilings and snow dumpings;
- Borrow pits to hold water for irrigation or ornamental ponds; and
- Traffic monitoring stations.

Industrial, commercial, and motor freight transporter operations might find it expeditious to lease portions of the highway right-of-way:

- Scientific research laboratories;
- Recycling depots (aluminum, paper, plastic);
- Truck terminals and offices;
- Maintenance shops; and
- Loading docks.

Organizations with a pastoral bent may be interested in:

- Zoological exhibitions;
- Botanical gardens;
- Fish hatcheries;
- Botanical nurseries (for future transplantation of young stock to other locations on the right-of-way or surrounding areas); and
- Agricultural research centers.

As already suggested, use of highway rights-of-way for other transportation-related activities has considerable merit:

- Public transit (passenger stations and use of median strips);
- Commuter park-and-ride lots;
- Bus depots beneath elevated highways;
- Storage for equipment and supplies; and
- Railway corridor.

Smaller urban remnant parcels can be used in myriads of ways for recreational pursuits, either as part of other existing or proposed parks or as distinct entities by themselves:

- Regional parks may be conceived to include ball diamonds, playground equipment, ponds and open spaces, archery ranges, off-the-road motorbike trails, bicycle paths, bridle trails, picnic tables, obstacle course–physical fitness centers, fishing ponds, floating docks–fishing piers, and boat launching ramps.
- Local parks may encompass playgrounds, play fields, indoor recreation centers, ice skating, and roller skating.
- Ornamental parks with creative landscaping, cultural displays, and fountain areas can provide aesthetic amenities.
- Under-structure possibilities, such as swimming pools, skating rinks, sport dog training grounds, and police pistol ranges, may not require too much noise attenuation from highway noise.

Rest areas are already included on major freeways, expressways, and arterials. Construction of new rest areas or expansion of view areas may consider including:

- Restrooms;
- Scenic turnouts;
- Information centers;
- Picnicking;
- Overnight camping;
- Historical views;
- Archeological salvages;
- Snack bars;

**FIGURE 3-2.** Expansion of existing rest and view areas may lead to inclusion of information centers and picnic tables in comfortable settings.

- Services;
- Freeway conveniences, including restaurants, grocery marts, motels, information booths, and vending machines; and
- Community services encompassing parking, gasoline stations, post offices, pedestrian overpasses, and enclosed storage of impounded and abandoned vehicles.

In addition to utility corridors, multiple use may also embrace:

- Water quality control facilities;
- Natural gas regulator stations and transmission pipelines;
- Electric sub-stations, overhead power, and conduits; and
- Sewage and waste disposal plants and/or pipes.

### Criteria, Caveats, and Cautions

From the preceding list, it can be readily appreciated that urban, suburban, and rural areas require different measures for applying multiple use concepts. Urban and suburban projects would involve people-oriented activities, along with commercial endeavors and economic generators. Rural projects would counter with land-oriented involvement to a considerably greater extent.

Existing highways ought not to be ignored should there be some glaring omissions in productive land use. Large interchanges can be ready targets for introducing appropriate multiple uses. Traffic and safety issues must be examined in each case, but, as described above, activities that could thrive in a restricted access situation may well put presently unused and difficult-to-maintain land into productive use.

Potential changes in the neighborhood, often triggered by a new or modified highway, ought to be contemplated and pondered. Maybe a specific multiple use

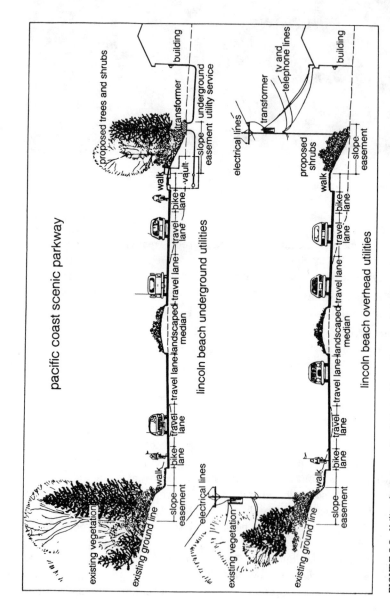

**FIGURE 3-3.** Utility sharing permits placing cables underground, thereby enhancing aesthetics and presenting opportunities to extend roadside landscaping. (Cross section rendering: Oregon Department of Transportation.)

would not be such a good idea if it would cause future disruption to the normal or planned changes in land use that result from improvement in the transportation network.

Maintainability of whatever multiple use is proposed is another ingredient to be thrown into the mix. This brings us around to determining how and what access needs to be provided in order to maintain any multiple use activities without jeopardizing highway operations.

## JOINT DEVELOPMENT

The next step in exploring multiple use of highway right-of-way is to determine if (and who) has a willingness and/or desire to jointly develop multiple uses. Often, this approach can aid—to some extent—the financing of a transportation improvement. Joint development projects have been carried on by highway agencies for decades, often without being designated as such—for example, utility sharing agreements have been established prior to construction, and beautification projects coordinated with adjacent communities have taken place in urbanized settings. Therefore, it can be seen that joint development involves cooperating with private entrepreneurs, corporate ventures, and various governmental groups on the national, state, or local level.

Investigating, instigating, and finally developing the various multiple usages that would be applicable to particular areas under consideration can result in considerable benefits. Policies to involve private businesses and civic-minded groups are being encouraged more and more. Further, it should be pointed out that agency participation can be greatly enhanced by projects that bring tangible benefits to the agency or community they serve. Summaries on special assessment impact fees and development agreements (from a legal perspective) are contained in *NCHRP Digest 161*, pages 8, 15, and 22.

Initial impetus for pursuing multiple use-joint development projects may come very early during the planning phase or not until late in the design phase when acquiring land for highway right-of-way. Parcels are often broken up or severed from the main body. When this occurs, the remaining pieces of land may become landlocked or they may be too small to be of great value. Rather than render such pieces of land useless and of little value, the highway agency (where permitted by law and policy) acquires the whole parcel, which is then sold to adjacent property owners or as surplus property, or is possibly used for development in one or more of the following ways:

- The agency's own use (maintenance stations, etc.);
- Joint development projects (on a lease or easement basis for public or private developments); and
- Agreements for multiple use (not necessarily associated with extensive construction).

Nevertheless, care needs to be taken so as not to plan joint development projects where constraints imposed by right-of-way use prohibit future upgrading of the facility. (For example, a building cantilevering over a freeway could prohibit vertical and horizontal expansion of the road section.)

Since the highway agency has jurisdiction over its right-of-way, joint development should in no way conflict with the needs or safety factors of the highway and its users. Proposals for joint development must be compatible with and should be complementary to highway use.

Where it becomes necessary to obtain land for highway right-of-way through highly developed commercial or industrial property, it may be economically advantageous to purchase "air rights." For example, by constructing an elevated roadway, the highway agency could decrease expensive right-of-way costs, and the land below could still be put to profitable use by private enterprise (which would retain title to the land). This aspect of joint development has yet to achieve a high degree of favor or acceptance.

### Right-of-Way Control and Protection

Joint development concepts open up the battery of dilemmas that control and protection of the highway right-of-way entail: a firm array of reasons why highway agencies require extensive coordination, documentation, review, and so forth, before committing to most of the previously cited opportunities for multiple use and joint development. Some "philosophical" points that the architect might keep in mind include procedures for controlling right-of-way to provide for safe and efficient operation of highways so as to use the full potential of the highway investment.

### Early Agreement

As soon as it is feasible in the development of a highway project, the necessary joint development agreements need to be initiated—especially those describing maintenance responsibilities. Waiting until the project is ready for construction could void all the efforts to effect multiple uses.

Land banking, concessions, and donation of land from entrepreneurs who stand to gain from proximity to a highway are three other realms that need to be coordinated at the early stages of design.

**References**

AASHTO. *A Guide for Transportation Landscape and Environmental Design*. 1991. Washington, D.C.: American Association of State Highway and Transportation Officials.

Bowman, B. L., J. J. Fruin, and C. V. Zegeer. 1989. *Handbook on Planning, Design, and*

*Maintenance of Pedestrian Facilities, FHWA-IP-88-019.* McLean, Virginia: Federal Highway Administration.

King, G. F. 1989. *NCHRP Report 324. Evaluation of Safety Roadside Rest Areas.* Washington, D.C.: Transportation Research Board, National Research Council.

Martz, Jon W. 1988. Infrastructure financing and joint development. In *Transit, Land Use & Urban Form*, ed. Wayne Attoe, pp. 161-170. Austin, Texas: Center for the Study of American Architecture.

NCHRP Digest 161. 1987. *Public and Private Partnerships for Financing Highway Improvements.* Washington, D.C.: Transportation Research Board, National Research Council.

Transportation Research Record 913, 1983. *Roadside Vegetation, Restoration and Protection.* Washington, D.C.: Transportation Research Board, National Research Council.

# Part 2

Preliminary Design

# 4
# Mapping

Some highway design offices, as soon as they receive their charge from the sponsoring agency, proceed directly into the task of coming up with a *final design*, even on complicated major projects. All too often, such an approach can lead to inordinate delays in completing the task. *Preliminary design*, if properly executed, can effectively shorten the time frame for the entire design process. The reason is that the myriad of trade-offs involved can be explored, prioritized, evaluated, and weighed before a great deal of effort is expended in going down pathways that lead to dead ends.

Many highway agencies refer to preliminary design as the "location phase." For this reason, "location studies" is occasionally used as a surrogate for preliminary design. And the first thing preliminary design must deal with is the topography, which includes streams, lakes, mountains, vegetation, and major structural edifices (together with their configuration and relative positioning); each are a part of the total topographical setting (see Fig. 4-1). Additionally, because of its expected longevity—perhaps a century or more—a highway alignment also becomes a part of the total topographical setting. Early in the design, the highway architect must become familiar with the topography to be traversed by a proposed roadway improvement concept. Three gambits to open play are:

- Surveying;
- Plotting cartographic features; and
- Reconnoitering.

36  Highways: An Architectural Approach

**FIGURE 4-1.** Mapping of topography includes structures and fence lines as well as major landforms and waterways.

## SURVEYING

Very rarely does one find a party in the field doing a "topo" survey. Field parties are most often now employed to set controls for aerial photography, to establish baselines through heavily wooded areas, to define exact positions of items to be incorporated into final design plans, and to locate points and elevations during construction. Preliminary field surveying has been superseded by mapping and by obtaining information already collected for other purposes. Searching out and pulling together existing surveys of water-related and geologic elements can be a rewarding adventure if pursued diligently.

### Waterways, Wetlands, and Floodplains

Each state has an agency (Water Rights, Riparian Bureau, Hydrologic Resources, or some similar title) that measures and records river and major stream flows at critical locations. In the federal government, the USGS, the Army Corps of Engineers, and the Bureau of Reclamation have data that could be applicable to a specific waterway. On major federal land holdings, a land management agency (Forest Service, Bureau of Land Management, National Park Service, or Bureau of Indian Affairs) may have useful information on stream flows. Judgment calls often come into play in cases where different agencies provide varying data pertaining to the same stream—something akin to the dispatcher who wears two wristwatches and is therefore never exactly sure just what time the train left the yard.

High quality maps showing floodplains are available for virtually the entire country through the Federal Emergency Management Agency (FEMA). Managed coastal zones are under the jurisdiction of individual state coastal zone management programs (CZMP). Since each program must be approved by the U.S. Department of Commerce, guidance on how and where to contact the applicable CZMP (if one pertains) can be solicited from a local office of the National Oceanic and Atmospheric Administration in the U.S. Department of Commerce.

## Geologic Hazards

A state land grant university generally acts as a repository for geologic records. The highway architect would be most interested in pinpointing potentially unstable ground: landslide susceptible areas, older landslide masses, and active landslide masses; debris flows and areas with potential for debris flows; land mass creep; rock falls; fault lines; avalanche hazards; problem soil units; and areas subject to flooding or high ground water problems. The relative ranking of the severity of each hazard can be shown or coded; it is often impossible to avoid all geologic hazards. Eventually in the design process, some ranking method would be desirable when assessing trade-offs involved with establishing roadway geometry.

## Controls and Baselines

Setting a baseline for any section of new alignment, to be used as a reference, is usually done at the beginning of preliminary design. Likewise, horizontal and vertical controls (coordinate points and elevations) are established concurrently with the baseline. More details on these procedures can be found in a handbook or text on route surveying.

## State Plane Coordinate System

Since many surveys are conducted at elevations somewhat higher than sea level, datum adjustments are made to ground level measurements in hilly country, high plains, and plateaus. Almost all states have a coordinate adjustment system. Some use the *Lambert Conformal Conic* projections, others employ the *Transverse Mercator Cylindrical* projections. Although the architect need not become expert at datum adjustment, an awareness of its use can help in communicating with surveyors. A detailed description (actually the "bible") of the plane coordinate systems can be found on pages 51–81 of *Highway Research Record 201*.

## Global Positioning System (GPS)

In surveyor circles, GPS does not stand for Gallons Per Second or Gol-dang Postal Service. The *Navigation Satellite Timing and Ranging System* (NAVSTAR), commonly known as *Global Positioning System* (GPS), is a space satellite chain developed by the Department of Defense. A 24-satellite constellation is to be in full operation by the mid 1990s. They will provide precise three-dimensional information on a continuous basis, which will supplant traditional field control surveys. Accuracy and lower projected costs are the rationales for the survey community going to GPS. When the full system is in operation, all survey controls will be referenced to GPS. More on the concepts of the subject can be gleaned from pages 7-4 and 7-5 of *FHWA/OH-87/010*.

### Geographic Information Systems (GIS)

Computerized mapping, combined with a completely managed database, is the up-and-coming method of plotting topography. An available extensive database, together with expensive software, are required. The kicker is having available an extensive database. Agency-wide, spatially related databases are not yet commonplace. Although all major highway agencies use some form of a referencing system, it more often than not is inconsistent throughout the agency. Files for different data groups were most likely created independently of one another, using different computer formats and reference bases.

Some digital files are available through the United States Geological Survey and the Bureau of the Census. The former sports a *National Digital Cartographic Data Base* (NDCDB) for the contiguous 48 states, embracing geographic references, boundaries, culture sites, federal and state land ownership, geodetic control, hydrography, hypsography, transportation, vegetative cover, and other miscellanea. The Bureau of the Census has been building a ferocious topological data bank of all the streets and blocks in the country, which they aptly refer to as the *TIGER* file. This brings us to our second gambit.

## PLOTTING CARTOGRAPHIC FEATURES

A highway on new alignment will find this aspect of cataloging the topography to be absolutely essential. Even highway improvements on existing alignment contain sufficient modifications, such as widening, additional taking of right-of-way, and interchanging, to require at least a modicum of cartographic effort. Special purpose maps of the surrounding area can be complementary (property tax maps, land use maps, soils maps, relief maps, geologic maps, and others most likely collected during planning efforts). Maps should deliver a good comparative picture of the environs through which the prospective highway alignment (including alternatives, as appropriate) could traverse. However, an excessive quantity of detail on a single map may render it less than useless—appearing as if the alphabet soup got strangled in the blender with a pot of half-cooked spaghetti. Information placed on a highway location map must be selective.

Mapping features initially of most value to the highway designer include:

- Scale and direction (north arrow);
- Elevations and/or contour lines;
- Waterways;
- Wetlands;
- Geologic hazards;
- Coordinate lines;
- Municipal and district boundaries;
- Cemeteries;
- Edifices (buildings, houses, barns);
- Existing roads, trails, and pathways;
- Utility lines (above and below ground); and
- Railroads.

As a preliminary location study proceeds, a designer normally adds additional information dealing with environmental concerns, often by means of different color or by clear film overlays. Computer-aided drafting and design (CADD) can be applied either way. Basic environmental concerns not included above embrace:

- Ecologically sensitive locales;
- Wooded areas; and
- Major classes of land use (farms, airports, and industrial sites, residential and commercial zones).

## Contour Lines and Their Characteristics

Although most aerial orthophotographs and mosaics come equipped with contour lines, there are cases in which contour lines must be placed on highway location maps from spot elevations, occasionally supplemented by rough contours. Interpolating contour lines is best done by photogrammetric means. Where this is impractical or not cost-justified, the use of appropriate calculator or computer programs may be employed.

The vertical distance chosen between contours (the contour interval) depends on the scale of the map and on the character of terrain. For larger scale maps of flat country, the interval may be as small as 1 foot; for smaller scale maps of rough country, the interval may be 40 feet. For maps of intermediate scale, such as those used for location studies, the interval is usually 2, 5, or 10 feet.

Basic principles for reading contours:

- Since each contour line follows a specified ground elevation, it will not merge or cross another (except in cases of overhanging rocks).
- Areas within contours are equivalent to "islands" above sea level; therefore,

contour lines always close (sometimes outside the borders of the map). A contour line containing no other contours within its closed area indicates either a summit or a depression. If elevations of adjacent contour lines or aquatic boundaries do not show which applies, a depression may be labeled using a hatched area within the contour line.
- A single contour line had better not lie between two contour lines of higher (or lower) elevation.
- Inasmuch as contours represent level lines, they are perpendicular to steep slopes as well as to ridges or valleys.
- Horizontal distance between contour lines becomes inversely proportional to the slope so that steep slopes cause contour lines to be closer together.
- On uniform slopes, contour lines space themselves uniformly.
- Throughout plane surfaces, contour lines tend to be straight and parallel.

**Aerial Photographs**

Quite a few of the maps for highway location studies are developed from controlled aerial photographs. Yet frequently, in the interests of economy, a proposed project would not be "flown" until after completion of preliminary design, a stage when the alignment is pretty well locked in. In such cases, previously flown photography, often taken for some other purpose, may be available and suitable for use in working up location study alternatives. In many instances, highway location terrain has already been flown for:

- Mining companies;
- Industrial developers;
- Real estate speculators;
- Planning and zoning departments of municipal governments;
- Major earth moving contractors;
- Water resource specialists;
- Geologists; and
- University geography departments.

If it is possible to latch onto any of these information sources, either directly or through aerial contractors who performed the work, considerable savings in flight-time expense can be had. Discrepancies in scale can be readily corrected by enlarging or reducing the photos. Another advantage of acquiring existing photography is that in most areas either vegetation or snow cover disguises the true nature of the ground through much of the year; previously flown photography was probably done during the season when these conditions did not pose problems.

## Orthophotographs and Mosaics

Extreme accuracy with regard to drainage, identification of soils, earthwork, material sources, and existing pavement condition is not always necessary during location studies and preliminary design. Nonetheless, if readily available, orthophotographs or mosaics or both remain the best regimen to follow for early mapping. An orthophotograph looks like a single photograph, although made of many matched aerials that have been adjusted for slant and scale. It will show a greater array of details than prosaic mapping. Uncorrected aerial photos joined to form a mosaic (considerably less expensive than orthos) are usually satisfactory for making presentations to civic groups and reviewing personnel. On the other hand, orthophotographs are more adequate for purposes where precision in identification is a prime requisite.

## Overlays

Situations continually arise in which alternative highway locations bump into other basic land uses. Existing and projected developments, public recreation areas, scenic vistas, historical edifices, unsatisfactory geological conditions, plus abundant additional constraints, may require that each theme be drawn on its own electronic layer, physically equivalent to creating a set of transparent overlays. One by one, such layers or overlays can show effects of the sundry land uses. Where an adequate database is available, GIS can fit into this function very well. Exhibits of this nature may be helpful, not only in studying trade-offs related to design, but also, as with mosaics, in describing alignment alternatives to interested community organizations and individuals.

## Photogrammetry

Photogrammetry based on aerial photographs is now a basic tool for use by all highway agencies. Aerial photographs taken with the camera pointed at the topo directly below are the convention for highway mapping. The immediate locale is shot in successive photos in parallel flights so that individual pictures overlap in both the flight direction and the parallel runs.

Photogrammetry has become more and more sophisticated than it was when partially controlled black-and-white photos were employed to obtain the "bird's-eye view." Color film with good resolution, stability, and sensitivity, along with distortion-free, color-corrected camera lenses give the highway architect a better mechanism than black-and-white photographs. More suitable information can be obtained regarding distinct land uses, surface conditions, borrow sources, water flows, geology, and drainage patterns. Color infrared film and radar sensoring can additionally complement color photogrammetry when a project is of sufficient complexity to justify the expense.

LANDSAT information, as it becomes available, digitized, and cost-effective, will probably supersede aerial photography.

**Computers**

Functionality in mapping is constantly being enhanced by prudent use of computers. Computers operating in association with stereo-plotters for map making during location studies, along with computer-based graphics, are more and more being used to bring actual on-site conditions into the architect's "studio" and then onto the drawing board (mouse pad). Computer graphics may include graphs, numbers, words, schematic displays, and even pictures. When the display shown is not quite that which is desired, the designer can input revised data, push a few keys and observe the resultant changes.

It is now a given that all architectural and engineering consultants and highway agencies have computer access for analyses and data processing vital to many aspects of highway design. Most administrators recognize that routine data

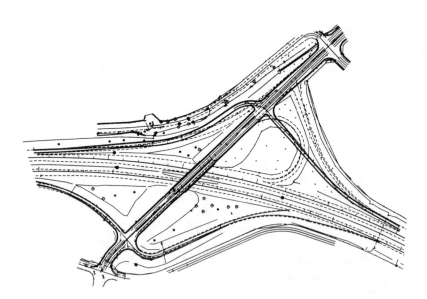

**FIGURE 4-2.** Computer graphics can aid both designers and reviewers. The plan view of a proposed interchange modification, above. Existing earth form (sans structure), right. (Graphics: Utah Department of Transportation.)

processing and reporting are best done almost exclusively by computers. Much nonroutine data manipulation is also appropriate for computers. The role of computer graphics (combining computer-generated data and analytical programs with a display and a graphical printout), together with remote sensing (the use of aerial photographs, radar, and ancillary devices for map making) are able to provide successive levels of preliminary mapping. Computer-recorded data is quite useful to generate separate maps that display unique informational elements. Drainage structures, utility lines, various land uses, zoning, existing roads, and lots of other pieces of useful (and, at times, useless) information find their way onto maps in this manner.

**Beware of Overkill**

Elaborate refinements tend to proliferate themselves. Photogrammetrics, combined with computer and digitizer, can furnish digitized topographic models, allowing information to be recalled and programs to be developed for displaying cross sections, profiles, embankments, and views of and from the road.

Before one gets too carried away, it is necessary to point out a major difficulty with rampant and reckless use of computers: dependence on computer printouts can waylay unprepared or lethargic designers who assume that certain often-used programs and inputs are doctrine. A dangerous tendency of the times arises when blind faith replaces technical judgment, resulting in the use of software of limited applicability, inappropriate computer programs, and garbage laden printouts. Adopting results from this process can frequently result in a badly tainted design effort.

During preliminary design, too much use of computers as a surrogate for value judgments is best avoided—even if it means performing manually some operations

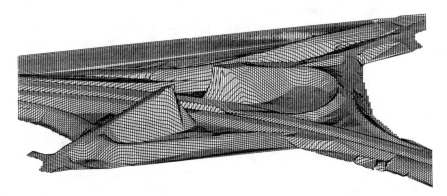

FIGURE 4-3.

that are considerably more efficient when computer-generated. Further on in project development, when design standards enter the picture, the balance between educated subjective judgment and computer efficiency drifts towards the latter.

## Map Content

Major highway improvements require a rather sophisticated and complete mapping effort. Minor improvements may not. (Should a proposed highway improvement be no more than a small widening on existing alignment, location mapping and much of the preliminary design effort need not be redone—assuming that the existing geometric conditions, for the most part, remain satisfactory.) We'll prescribe the following for a major mapping effort.

*Scale and direction*—Map scale for preliminary work may range from 1 inch = 100 feet for complex urban projects to 1 inch = 2,000 feet for projects of a rural nature. (The latter corresponds to the scale used on most older United States Geologic Survey topographical maps; the new ones are being done on a 1:25,000 metric base.) 1 inch = 400 feet is probably the most commonly used scale for location and preliminary design work. And yes, a north arrow.

*Elevations and/or contour lines*—Contour lines are preferable where a project is to be in rolling or mountainous country; contour intervals need to be appropriate to the conditions so that there are an adequate number of lines to describe the ground surface and to interpolate from, yet not so many as to create a "busy" map. Spot elevations (instead of contour lines) are better suited to urban projects and those on very flat surfaces.

*Waterways*—Initially, it is advisable to plot shorelines of oceans, gulfs, bays, rivers, and lakes, plus ponds, canals, creeks, streams, brooks, washes, draws, and irrigation ditches, if they are in the reasonable vicinity of the alternative alignments being considered. The extent of tides, floodplains, estimated or recorded flows (including direction), depths, seasonal fluctuations (where relevant), and even drainage basin areas are valuable adjuncts to preliminary maps, provided they are discreetly placed. Water can be an enemy or it can be an ally in designing, constructing, operating, and maintaining highways. A good deal of care to map properly all forms of waterways is essential in the early stages of highway location work.

*Wetlands*—Swamps, marshes, bogs, and managed coastal zones, for the most part, are given general boundaries where no sharp lines of demarcation exist. Wetlands, though of considerably more than casual concern in highway design, have an even greater import as habitat for many forms of wildlife. Across the country, some different concepts are applied to swamp and marsh. Swamp can generally be considered as land that is covered much of the time by shallow

**FIGURE 4-4.** Boundaries of waterways and wetlands that continually vary and meander make accurate mapping somewhat problematical.

water; marsh would be wet land full of cattails or long grasses or both. There need not be a fussy distinction between the two on the part of the highway architect; they both pose location problems of a similar nature.

*Coordinate lines*—Use of the State Plane Coordinate System will initially provide a smoother transition from preliminary to final plans.

*Municipal and district boundaries*—A hierarchal scheme of long dashes, short dashes, and dots should be employed in delineating boundaries of cities, counties, water service areas, fire districts, and so on.

*Cemeteries*—For the longest time through recorded history, the principal curse placed upon the highway designer has been the presence of cemeteries directly in the path of the best highway alignment. All cemeteries and graveyards need to be "plotted." It is recommended—rather strongly—that a proposed highway re-alignment not go through a cemetery or graveyard.

*Edifices*—Many veterans of the military service are familiar with the basic instructions: "If it moves, salute it; if it doesn't move, move it; if it's too big to move, paint it." People-made objects that, according to this directive, should be painted are usually of sufficient magnitude and importance to be plotted on the map and labeled.

*Existing roads, trails, and pathways*—Items of this nature are considered essential to the location studies; they too should be located (and labeled) on the preliminary map.

*Utility lines*—The same directive applies to gas, water, electric, sewer, and pipelines, as well as their major terminals and junction points (whether above—or below ground level).

*Railroads*—Rail lines are important guides in determining where the lower gradients can be found. In fact, for any particular project, a railroad may have already usurped some of the best potential highway location. (A railroad surrounded by historic graveyards can be a real bear should the architect find such a corridor to be the only suitable alignment location for a proposed highway.)

*Ecologically sensitive locales, wooded areas, and major classes of land use*—"Discretion is the better part of valor" when determining the extent of additional information to be placed on maps to be used for location studies and preliminary design.

## RECONNOITERING

If at all possible, the architect should pay a visit to and traverse the locale of the proposed new or remodeled highway. Taking along the preliminary maps or aerial photos and a camera (or VCR), the architect's initial visit to the site can quickly enhance an understanding of the topography. Ground level photos may be used later to aid in drainage design, earthwork estimates, alignment considerations and right-of-way determinations. Marking directly on the map or aerial the precise location (and facing direction) where each photo is taken can relieve a lot of anxiety some months into the future when recollections become a little hazy as to what each photo illustrates. Many times, more than one visit to the project location is justified, more often when each trip has a singular purpose.

### Underground Utility Lines

In some jurisdictions, utility companies maintain excellent records describing just where and how deep their underground pipes and conduits lay. In all too many cases, though, records of this nature are:

- Inaccurate;
- Incomplete;
- Outdated;
- Non-existent; or
- All of the above. (How's that again?)

On-site reconnaissance may be the only way to ascertain underground utility locations in the vicinity of proposed and existing highways. A crowbar is a useful tool to take along on the reconnaissance in order to dislodge stubborn manhole covers. Spotting the positions of water valves, peering through catch basin grates for sewer pipes inlets and outlets, inspecting contents of junction boxes (very carefully if they happen to belong to the electric or gas companies), and observing other telltale signs that indicate there may be nearby underground utility lines is the maximum extent usually given to this subject during location studies. A more detailed investigation is required during final design activities.

## Existing Drainage

To observe the condition and general nature of drainage patterns and facilities is another good reason to visit the site of the proposed roadway improvement.

## FULFILLMENT

Completion of this activity should find the architect's work station inundated with sufficient maps, photographs, references, floppy discs, and sundry ancillary paraphernalia to facilitate getting on with actual preliminary design. Before proceeding, however, it is mandatory that each piece of evidence collected be properly labeled and, for the more complicated projects, cataloged.

### References

*Global Positioning System (GPS) Exploitation by Ohio Department of Transportation*, Report No. FHWA/OH-87/010. 1987. Washington, D.C.: Federal Highway Administration.

NCHRP Research Results Digest Number 180, *Implementation of Geographic Information Systems (GIS) in State DOTs*. 1991. Washington, D.C.: Transportation Research Board, National Research Council.

O'Neill, Wende A., and Balakrishna Akundi. 1990. Automated conversion of milepost data to intersection/link network structures: an application of GIS in transportation. *Transportation Research Record 1261*:27-34. Washington, D.C.: Transportation Research Board, National Research Council.

Pryor, William T. 1967. Precision in surveys by use of plane coordinates. *Highway Research Record* 201:51-81. Washington, D.C.: Transportation Research Board, National Research Council.

# 5

# Construction Materials and Foundations

There are probably over a thousand different materials that go into a typical highway project. Here we are only going to touch on those few materials that comprise the bulk of the costs in a construction contract, namely:

- Soils;
- Bituminous products;
- Portland cement;
- Metals (structural and reinforcing); and
- Geosynthetics.

Corollary to extracting materials are the foundations provided by soils and selected materials.

## SUBSURFACE

Raw materials taken directly out of the ground are sorted and remixed to form about 90 percent of the materials in the highways we drive on. Not to be ignored are the soils that are left in place to support the pavement and appurtenant structures. A project that is well thought out, therefore, should have considerable developmental effort spent in determining optimum use of these raw materials.

Unfortunately, some of our major highway agencies abdicate their responsibility for accurate and thorough subsurface investigations, pushing it onto contractors bidding the project. A potential bidder may have a month or so to study and learn about the materials available before the bid is due. What if the ground is covered with snow, or if right-of-way acquisition details have not been cleared before the bidding period, or if environmental restrictions prevent subsurface investigations by nonpublic entities? Many times there is no way for a likely bidder to know what is underground when referring to agency subsurface investigations.

So we have a dilemma that can confront a highway architect early in project development. Can the policy of the highway agency (or entrepreneur if private capital is involved in a proposed project) be made less draconian? After all, most bidders will either tack on a contingency in their bid prices to cover the unforeseen, or alert their legal counsels to prepare for an expected court skirmish should unexpected subsurface conditions cause big problems.

The presumption will be made that the architect will see to it that adequate preconstruction investigations are made and that the sponsoring highway agency will assume responsibility for the reasonable representations required. This assumption applies not only to the quality of what lies underground, but also to the quantities and availability—matters that are further alluded to in subsequent chapters on water, earthwork, and drainage.

## SOILS AND GEOTECHNICS

Soil (dirt and rocks to those of us who tend to be uncivilized) forms two important components of highways:

1. Support for pavements and structures;
2. Materials for building pavements and structures.

Soil contains not only solids, but liquid and gas as well. The solid portion is comprised of various size particles arranged in randomly diversified distribution; the liquid portion is mainly water filling some of the voids between solid particles; the gaseous portion (air) occupies voids not filled by liquid. All three play critical roles in highway and structure design.

In order to identify soil properties in any specific location, as already stated, subsurface investigations are requisite. They can be subdivided into four steps, each of which ought to supplement information from the preceding step. Essentially, these steps are:

1. Armchairing;
2. Sampling;
3. Testing; and
4. Interpretation.

Some projects can be worked up without going into steps 2 and 3. However after whatever level of investigation (or exploration) is completed, a competent-level "Interpretation" must be undertaken. It just doesn't do justice to the overall design effort if only the boring logs are included on the plans, without any annotation relating the subsurface investigation to supportive ability or materials use.

**Armchairing**

Most sites where new or improved highways are to be located have already had considerable soils investigation performed—often for other purposes. Also, the sundries already gathered, such as aerial photos (particularly infrared), USGS quad-sheets, and topographic maps, can help provide an overview. It would be prudent, as circumstances dictate, to contact the following information sources for available soils and geologic information:

- Agricultural maps (Soil Conservation Service);
- Oil, gas, and water logs (local well drillers);
- University geography departments; and
- Local construction history.

Low-budget projects often get only a cursory type of soil investigation. Larger budgets and the more critical projects require a considerably greater effort to obtain subsurface information. Sampling and testing would be the next steps for the more formidable undertakings. So the question arises: Is there a handy rule of thumb that could tell the architect just how much subsoil should be checked out?

Answer: No.

**Sampling**

Obtaining soil samples is subject to certain protocols. Categorizing the various methods gives us:

- Auger boring;
- Displacement boring;
- Wash boring;
- Percussion boring; and
- Rotary drilling.

Each of these has its place in the hierarchy of investigative trade-offs.

The ultimate objective of boring (and drilling) is to obtain information on subsurface conditions, not just to punch holes at the site. Investigations should provide much of the following:

- Water levels;
- Depth and thickness of each stratum;
- Permeability;

- Field tests of other soil parameters; and
- Samples for determining soil properties in the laboratory.

Number and spacing of borings are usually based on educated appraisals (guesses), which are then modified according to field conditions. Depth of borings often follows the common practice handed down from experiments done at the Geotechnical Institute of Belgium; the designer may have to figure out the stress distribution from which an estimate can be made of the depths for the borings. Without going to a more sophisticated approach, some orthodox recommendations for spacing and depth of borings are suggested in Table 5-1.

Planning the boring operations becomes mandatory when there is a large expense in moving to and from a site or in working at a location that is distant from the testing laboratory. (In such situations, it is often economical to perform additional—even redundant—borings or to procure extra samples than to return a rig to the site.)

Once obtained, samples can be divided into "disturbed" and "undisturbed." Disturbed samples are relatively complete, but contain significant structural derangement. They are suitable for classification tests, such as by penetration methods. Undisturbed samples, in which the structural disturbance has been kept to a minimum, are more difficult to obtain. Although impossible to extract a perfectly undisturbed sample, it is feasible to keep disturbance minimal with skillful and careful effort.

## Testing

Density, the most important soil physical characteristic in highway architecture, is primarily affected by moisture content. Laboratory tests set standards for density, while field tests measure density of soils in place. Supporting power (strength)

TABLE 5-1 Recommendations for Spacing and Depth of Borings

| Boring For | Spacing | Depth |
|---|---|---|
| Roadway Profile | 1,000 ft. | 6 ft. below subgrade elevation |
| Cuts | 4 points along a straight line perpendicular to centerline through extent of cut | Depth of cut |
| Fills | Same as for cuts | 2 times height of embankment |
| Structures | 3 minimum (not on a straight line) | Educated guess based upon weight of structure |

under one moisture condition may become quite different should the percentage of moisture change—and soils supporting subgrades are constantly subject to moisture changes. A 1-percent variation (measured as part of the total weight) in the moisture content of granular layers may cause pavement damage to increase by a very significant margin.

Although some soils remain at a constant value, there are many that, when compacted at optimum moisture content, will swell when added moisture becomes available. Swelling may not occur if the external pressure on the soil mass is great enough, so we might expect that soil that is buried deep will not usually cause trouble. In cases where confinement under high fill is imprudent, volume changes can be reduced by compacting the soil at a higher moisture content so that the demand for water in the soil is somewhat assuaged. This clearly indicates that testing for optimum moisture content is a must for any soils that are to be used for supporting loads.

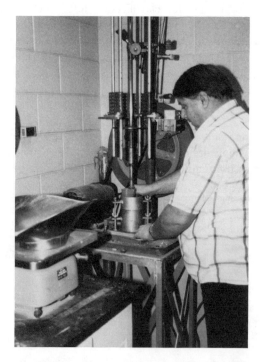

**FIGURE 5-1.** Optimum soil moisture is often determined in the laboratory by the Proctor Method, in which blows from a controlled hammer are applied to compact the layers of a confined soil sample. Various percentages of moisture are tested to ascertain the maximum density, a direct function of the optimum moisture content of the sample.

Classification tests, which determine the index properties of soil, include mechanical analysis, density, grain size, and sensitivity. Laboratory tests of this nature are undertaken so that the designer may draw conclusions regarding soil makeup, texture, and consistency. Knowledge of these properties allows for prediction of types and approximate magnitude of potential soil problems. In the appendix of the *Earth Manual*, the reader can learn more about how some of these tests are performed and what the testing equipment looks like.

It is becoming customary to perform on-site soil tests where feasible. Some of the common ones include:

- Penetration tests;
  - Standard;
  - Cone; and
  - Piezocone.
- Plate load test;
- Pressuremeter tests:
  - Full displacement;
  - Prebored; and
  - Self-boring.
- Dilatometer test.
- Spectral analysis of surface waves (seismic).
- Vane shear test.

Details of how these tests apply and what they show can be found in *Transportation Research Record 1235*.

Any differences between normally consolidated and over-consolidated soils ought to be reckoned with in order to decide on type and method of shear test. Normally consolidated soil is in equilibrium with its present overburden pressure and has not been stressed to a higher pressure than it is presently undergoing. An over-consolidated soil has been stressed sometime in its geologic past to a higher pressure than it is now experiencing from the overburden. An over-consolidated soil usually contains a natural moisture content well below its liquid limit, whereas a normally consolidated sedimentary soil is characterized by:

- Natural moisture content approximating its liquid limit;
- Increasing cohesion with depth; and
- Decreasing void ratio (along with concomitant lower water and gas content), with depth.

Effective stress determined from field and/or laboratory tests can aid in effective design, since it correlates directly with soil behavior. By increasing the effective stress of a soil, the particles come closer together with an added densification and an increase in shear strength.

## Geotechniques

Further field testing is quite often performed using geotechnical methods. Two such techniques have been in use for subsurface exploration: the *seismic* method, which uses the principle of travel speed of shock waves being different for different materials, and *resistivity*, where variation in the electrical conductance between different subsurface materials is correlated with known values.

The former (seismic) is used mostly for determining the depth to rock, although the method may obtain results that could be deceptive in stratified tilted materials. Since the equipment is lightweight and portable, it is suitable for most highway reconnaissance. The "sensitive" equipment required includes a sledgehammer and a small steel plate. *Refraction seismology*, as it is known in the trade, measures sound waves traveling through different materials at different rates, depending upon the degree of cementation and hardness or density. When these factors increase, so does the sound velocity. Refraction seismographs employ timing devices that measure the time it takes a sound wave to travel from the impact source to a not-altogether-fancy electronics gadget called a geophone (much like a microphone). Depth determinations can be made by graphs or formulas; however, each layer tested must be fairly thick.

The other approach (resistivity), using the principle that there is variation in the electrical conductance between different subsurface materials, requires a direct current to be passed through the soil between two electrodes. Any drop in potential is measured between two other electrodes placed intermediately at points one-third the distance from the actuating electrodes. The voltage drop measures resistance of the soil layer to a depth that is the same as the spacing of the intermediate electrodes. This method is used for locating sand and gravel deposits. Inasmuch as electrical resistivity is actually based upon electrical conductance (the reciprocal of resistivity), it is influenced by ions (dissolved salts), moisture content, and grain size. Saturated clay, thus, is a good conductor (low resistivity) and rock is a poor conductor (high resistivity). Most soils are between these two extremes; consequently, this method's merit lies in relating field results to actual physical features by producing calibration curves in areas where soil is exposed or near locations where the soil profile is known from boring logs.

The *AASHTO Manual on Subsurface Investigations* presents a more detailed treatise on sampling and testing for those architects who do not have a background in the subject and do not have ready access to a geotechnical specialist.

## Interpretation

At best, interpretation of soils data is an art form, with many built-in pitfalls and disappointments. Design, construction, and operational experiences have shown that, even when correct design and construction procedures are followed, early failures can occur after a wet spell—failures that are normally attributed to an unexpected increase in moisture content in the subgrade. Since moisture buildups in localized subgrades cannot always be predicted with accuracy, they are often blamed on changes in drainage patterns, topography, and other influences that take place during and after construction. Should the architect not be versed in the specialty, but likely to become involved in the interpretation of soils data, reference to a standard text on soil mechanics is highly recommended. If the architect needs to go further into the subject, a foundation design text or the *Manual on Design and Construction of Driven Pile Foundations* may be helpful.

## VALUE-ARCHITECTURE IN BASE CONSTRUCTION

A highway architect, faced with much uncertainty, may choose to cheat Mother Nature and attempt to evade some inherent problems with soil stability. One of the options gaining favor is the use of geo-textiles in roadway construction where soft, low-strength soil conditions prevail. Geo-textiles may be used in conjunction with readily available aggregates, such as crushed stone, gravel, or even sea shells, to develop a structural support layer. In situations where low-type surfaces are to be employed as planned stage construction for higher-type roads, experience has shown that geo-textiles can be cost-effective and may allow considerable lowering in the quantity (and sometimes even the quality) of aggregate used to help stabilize subgrades.

The use of existing embankments (as mentioned previously) may result in lower costs for a highway segment because they are already well consolidated. In situations where existing cut and fill sections have to be widened, some thought may be given to widening only on one side where feasible so as to minimize the manipulation of narrow cross sections. Topsoil and roadbed aggregate bases should be stockpiled when they can be reused effectively. An old roadbed may contain materials that would prove adequate for the base course of a paved shoulder. Costs involved may be considerably less than for a shoulder built with new base materials.

Any Value-Architecture connected with earthwork can do a great deal to help keep the overall cost of a typical project from getting out of control.

## SPECIAL AND UNUSUAL CASES

Sometimes the architect is forced to assume a lead role in assuring that unusual situations are given adequate attention. Soft foundation materials are one example. *NCHRP Synthesis 29*, pages 7-17, can provide some guidance in selecting and overseeing one or more treatment schemes. And there is always the specter of embankment failures during construction that need quick design fixes. *NCHRP Synthesis 33*, pages 16–18, presents an extreme example.

Something else to consider (which is not always an unusual or special case) crops up on a roadway cross section that is partly in fill and partly in cut. Support parameters are distinctly different for fill and cut. Thus, a homogenous subgrade is often problematical.

### Involvement of the Architect

Should the architect get heavily involved in geotechnical details because of unusual circumstances, he or she will quickly discover that the pedological soil classes shown in Table 5-2 may be useful for location purposes, but have little relevance to the classifications used during design. There must be more than a dozen classification methods in common use. Some of the classification schemes that a highway person is most likely to come across include:

*AASHO System*, which separates soils into particle size, plasticity, elasticity, and organic content;
*Textural System* of particle size distribution, in triangular charts;
*Unified Soil System* for use in subgrades;
*Engineering System*, which dwells on cohesion or lack thereof; and
*Muskeg System*, used in Canada and elsewhere for soils of high organic content.

The architect will be left to conclude that it's probably not necessary to learn them all.

### Phasing with Geometry

It would seem needless to point out that initial soils investigations must be done concurrently and in conjunction with preliminary determination of highway geometry.

## SOILS INVESTIGATION: FULFILLMENT

At the completion of a preliminary geotechnical/soils investigation, the architect should be aware of any problem areas regarding support for pavements and structures, locations of desirable borrow and/or disposal sites; recommended design of subbase, tentative inputs for pavement design, slope stability, depth to rock, limitations on type of drainage pipe, depths to groundwater and frostline, and compaction factors. (In many soils reports, there is a tendency to provide more information than needed.)

For all but the simplest projects, soils investigations are generally undertaken by a separate unit or division in a major highway agency, or by a consultant or laboratory that specializes in either geotechnics, geology, or soil mechanics.

## AGGREGATES

Sorting and remixing aggregates (derived mostly from soils) for use in base courses, asphaltic concrete, and portland cement concrete (PCC) is necessary (see Fig. 5-2). They are segregated into coarse aggregate, such as stones and gravel, fine aggregate (sand), and mineral filler. Where the aggregates are not cemented, they are usually structured so that there are sufficient voids to allow moisture to drain away. In concretes, the larger aggregates provide the strength and bulk, medium aggregates fill in the voids of the larger aggregates, fine aggregates fill in the voids of the medium aggregates, and the cements fill in that which is left and holds the entire mass together.

Gradation and hardness are two prime factors in the proper use of aggregates. These factors are initially determined in the design phase, but are more thoroughly examined and tested during construction when the actual material is uncovered, either from excavations on the construction site, quarries, or gravel pits or from commercial stockpiles.

Particle shape—sharpness—is important in concretes. Rounded aggregates such as beach sand are unsatisfactory because they do not lock onto each other to form a strong bond. Crushed (fractured) gravel is more satisfactory than round gravel. A movement is underway to use fractal geometry in determining characterization of aggregate shape (Carr, Norris, and Newcomb, 1990).

**FIGURE 5-2.** Aggregates are sorted into various sizes and then remixed into appropriate proportions that are suited to the use for which they are to be put.

TABLE 5-2   Capsule Guide to Soils for Highway Location

| Preference Hierarchy | Soil Class | Origin | Prominent Locations | Remarks |
|---|---|---|---|---|
| 1 | Kames & Eskers | Glacial | Alaska; Great Lakes States | Excellent sources of construction materials |
| 1 | Beach Ridges | Waterlaid | Great Lakes States; East & Gulf Coasts; Alaska | Very good road base and construction materials source |
| 2 | Schist | Metamorphic | Appalachia; Alaskan Range; Rocky Mountains; Northern New England; Lake Superior Vicinity | |
| 2 | Outwash | Glacial | Immediately to the south of terminal moraines | Excellent base for roads; good source of construction materials |
| 3 | Basalt | Igneous | Western States; Hawaii | Areas of tuff should be avoided |
| 3 | Coastal Plains | Waterlaid | Mid-Atlantic, South Atlantic, & Gulf States | |
| 4 | Sandstone | Sedimentary | Appalachia; Southern Great Plains; Rockies; Pacific Coast; Alaska | Often interbedded with shale; rock excavation tends to be costly |
| 4 | Granite | Igneous | Appalachia; New England; Adirondacks; Black Hills; Sierra Nevadas; Northern Cascades; Alaska | High excavation costs |
| 4 | Gneiss | Metamorphic | New England; Adirondacks; Alaskan Range; Rocky Mountains; Appalachia; Lake Superior Vicinity | |
| 4 | Alkali Flats | Waterlaid | Southwestern States | |
| 5 | Alluvial | Waterlaid | Mississippi Valley & westward to the Pacific | Coarse aggregates may be obtained from nearby alluvial fans |

| | | | | |
|---|---|---|---|---|
| 5 | Dunes | Windlaid | Seacoasts; Great Lakes States; Great Plains; Columbia River Basin; Texas–New Mexico; Alaskan Plains; Great Basin; Southern California | |
| 5 | Lake Beds | Glacial | Northern Great Plains; Great Lakes States; New England | |
| 5 | Limestone | Sedimentary | Appalachia; Southern Great Plains; Northern Rockies; Florida (coral); Hawaii (coral) | Often interbedded with shale; good source for surfacing aggregates |
| 5 | Till | Glacial | North of Ohio and Missouri Rivers; Northern Pennsylvania; New Jersey; New York; New England | Thin layers often cause drainage problems |
| 6 | Drumlins | Glacial | Great Lakes States; New England | |
| 6 | Loess | Windlaid | Mississippi-Missouri Valley; Pacific Northwest | |
| 6 | Moraine | Glacial | Alaska; Northern Great Plains; Midwest; Long Island; Southern New England | Frost heave often causes maintenance problems |
| 6 | Tidal Flats | Waterlaid | Near Seacoasts | Flexible pavements recommended |
| 7 | Flood Plains | Waterlaid | Near major rivers and flowing streams | Insurance costs often dictate against locating highways in flood plains |
| 7 | Slate | Metamorphic | Appalachia; Catskills | Often interbedded with anthracite; poor construction material |
| 8 | Deltas | Waterlaid | River mouths | Flexible pavements recommended |
| 8 | Shale | Sedimentary | Appalachia; Great Plains; Southern Rockies; Pacific Coast; Alaska | Often interbedded with sandstone and limestone; poor highway base |
| 9 | Organic Deposits | Waterlaid | Humid locales | Flexible pavements recommended where organic deposits cannot be avoided |

60     Highways: An Architectural Approach

## BITUMINOUS PRODUCTS

As a practical matter, the terms bituminous and asphalt can be interchanged with very little flack coming from members of the highway community. True, the term bituminous embraces tar and coal as well as asphalt. Asphalt is a more specialized product that now comes almost exclusively from the refining of petroleum.

In road building, asphalt is sorted into two types—cutbacks and emulsions. The former use diluents, such as naphtha, gasoline, or kerosene; the latter are dispersed in a mixture of water and emulsifier (soap or similar product). Further breakdown classifies cutbacks into rapid curing (RC), medium curing (MC), and slow curing (SC). To avoid confusion with cutbacks, emulsions are designated as rapid setting (RS), medium setting (MS), and slow setting (SS). And yes, the breakdown continues with a reference to viscosity (in the case of cutbacks). Thus, when someone drops the term "MC-800," he or she means a medium-curing liquid asphalt cutback with a kinematic

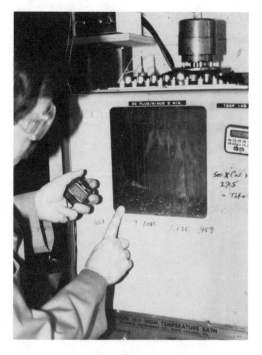

**FIGURE 5-3.** Laboratory measurement of an asphalt's absolute viscosity by a vacuum capillary viscometer: the time required for a fixed volume of the liquid asphalt (at 140 degrees F) to be drawn through a capillary tube (by means of a vacuum) multiplied by the viscometer calibration factor.

viscosity exceeding 800 (actually in the range of 800 to 1,600). Once one learns the jargon, the mystery in bituminous terminology disappears.

Emulsified asphalt is either anionic or cationic, depending on the emulsifying agent. Since anionic emulsions have a negative charge, they are more effective in coating electropositive materials, such as limestone. Cationic emulsions, with their positive charge, work better with siliceous materials, such as sand and gravel.

Emulsions, once felt to be problematic, have come into greater use since the 1970s and have replaced cutbacks in a variety of uses as a result of environmental concerns. Cutbacks tend to be viewed as pollutants because of the diluents used.

Those who work with asphalt and asphaltic products on a day-to-day basis have come to realize that theirs has yet to become an exact science. Recent findings indicate that part of the mystery of asphalt is its having both acidic and basic (amphoteric) properties. One authority on asphalt has stated (not surprisingly), "The chemistry of petroleum asphalt is very complex." Do we know, therefore, why asphalt pavements grow old?

> New asphalt pavement has a rich black look to it and has few surface defects. Soon though it begins to wear and age as vehicular traffic and the effects of sunlight, exposure to air, water, freeze-thaw cycles, etc. deteriorate the binder, or petroleum asphalt. As the petroleum asphalt begins to age and oxidize it loses its adhesive power, and the aggregate begins to work loose. Left unprotected, the asphalt pavement eventually deteriorates to the point of having to be removed and replaced. Analysis of asphalts from deteriorated pavements reveals the following:
>
> 1. They are low in penetration, and have very high viscosities. This indicates the asphalt has become hard and brittle.
> 2. They are higher in asphaltene and polar compounds content than the original asphalt. However, the asphaltene content is higher, proportionally, than the original, while polar compounds, proportionally, are about the same.
> 3. They are lower in aromatic and saturate content than the original asphalt.
>
> This evidence indicates that in the aging process, more asphaltenes are formed as the aromatic and saturate portions oxidize. Since no new polar compounds are formed, the excess asphaltenes are not in solution, and this causes the asphalt to become hard, brittle, and lose its adhesive power. (Partanen, 1988)

Grade and type of asphalt for each separate use is another of those art forms that derive their validity from experiences that include past successes and failures. Rarely is the highway architect or designer required to select the correct grades and types of asphalt employed in every aspect of a project. This matter has already been determined by the specifications and standards of the highway agency or entrepreneurial venture sponsoring the project. If there is an omission on their part, the Asphalt Institute is willing to come to the rescue with appropriate specifications for any specific situation.

Bituminous materials have other uses in addition to cementing aggregates, waterproofing being the most prominent.

## PORTLAND CEMENT

Several different types of portland cement are used in highway construction:

- Type I is the basic standard.
- Type II is sulfate resistant.
- Type III provides high early strength.
- Type IV has low heat of hydration.
- Type V has a high sulfate resistance.
- Type K has shrinkage compensating properties.

The most commonly used cement in construction is Type IIA. The letter A is often found following each type and designates an air entrainment additive to trap small, uniformly distributed bubbles of air when mixed into concrete. The purpose of air entrainment is to enhance durability.

Portland cement is also used in mortars for cementing bricks and stones. Often mixed with portland cement are other cementitious products: pozzolans such as flyash from coal-fired power plants (an effective mode of recycling), natural cement, or lime.

## METALS

Steel and aluminum are two metals that show up on major road building jobs. The former is used extensively for bridge girders, reinforcing in portland cement concrete, guardrails, and corrugated drainage pipes. Aluminum may be specified for railings, corrugated drainage pipes, and pipe arches (small bridges) where harsh climates or soils deteriorate steel.

## GEOSYNTHETICS

Several pages back, the use of geo-textile fabrics in conjunction with readily available aggregates was suggested. Other uses for geo-textiles, or geosynthetics, are continually being proffered. Most highway agencies now approve the stabilization of slopes and embankments, using geosynthetic reinforcement. The problems with geosynthetics are two fold: 1) specifications for one type of use may be entirely inappropriate for other uses and 2) they become easily damaged during their placement. An even, unbroken, smooth, uniform application of fabric on a hillside is difficult and problematical. The contractor may be faced with the same dilemmas that face the proverbial one-arm paperhanger—except on a grandiose scale.

## References

AASHTO *Manual on Subsurface Investigations*. 1978. Washington, D.C.: American Association of State Highway and Transportation Officials.

*Asphalt Handbook*. 1965. Lexington, Kentucky: The Asphalt Institute.

Bureau of Reclamation. 1974. *Earth Manual*. Washington, D.C.: U.S. Department of the Interior. (The appendix is contained in printings prior to 1990.)

Carr, James R., Gary M. Norris, and David E. Newcomb. 1990. Characterization of aggregate shape using fractal dimension. *Transportation Research Record 1278*:43-50.

*Design and Control of Concrete Mixtures*. 1979. Skokie, Illinois: Portland Cement Association.

FHWA *Geotextile Engineering Manual*, Report TS-86-203. 1986. Washington, D.C.: Federal Highway Administration.

FHWA *Manual on Design and Construction of Driven Pile Foundations*, Report DP-66-1. 1985. Washington, D.C.: Federal Highway Administration.

Hough, B. K. 1969. *Basic Soils Engineering*. New York: The Ronald Press Company.

Johnson, Stanley J. 1975. *NCHRP Synthesis 29, Treatment of Soft Foundations for Highway Embankments*. Washington, D.C.: Transportation Research Board, National Research Council.

Jones, G. D., Jr. 1976. *NCHRP Synthesis 33, Acquisition and Use of Geotechnical Information*. Washington, D.C.: Transportation Research Board, National Research Council.

Partanen, John E. 1988. *Asphalts and Agents for Recycling*. Unpublished paper.

Transportation Research Record 1235. 1989. *In Situ Testing of Soil Properties for Transportation*. Washington, D.C.: Transportation Research Board, National Research Council.

# 6
# Water

Fountains! Waterfalls! Lush vegetation! Exotic plantings! Sculptured shorelines! An architect is in celestial ecstasy when water becomes available to enhance a creative endeavor.

But we're talking highways. So it becomes necessary to realize that, in its many forms, water imposes a variety of impacts, challenges, and opportunities on how one locates and designs any highway segment. Rivers, lakes, streams, coastlines, wetlands, rainfall, snow, and ice, as well as irrigation for farms and water supply to communities, may all separately or in combination interact with the placement of or modification to any highway. This means that any dealings with water must be done concurrently with establishing or modifying highway geometry.

## LOCATING THE HIGHWAY

An ideal highway location, insofar as avoiding problems with water, would be aligned atop crests between major drainage basins. Attention then needs to be directed solely to water that falls directly onto the highway template. No problems with stream crossings, culverts, drainage ditches, and other such encumberances would present themselves. In addition, the vistas would probably be magnificent. Yes, but how about the highway user whose principal concern is to get from point A to point B in a reasonably efficient manner? A routing along the crestlines is satisfactory only in extremely rare situations.

Consider then the other extreme: a lowland roadway that closely parallels a river. The roadway must cross tributaries where they are the widest. Necessary structures, deep cuts, and massive embankments are costly to construct, and the latter two contribute to extensive troubles with erosion.

Thus, it appears that neither the high road nor the low road is going to get one to Scotland—or any other destination, for that matter—with any great efficiency. The selection of the best routing is an assembly of best trade-offs.

## Disruption of Waterways

Highway construction certainly has the potential to disrupt established water flow patterns (see Fig. 6-1). Roadway prisms, after all, do intercept water that previously travelled uninterrupted across a highway alignment. This factor requires strong consideration in any proposed roadway. Multi-water flows can usually be diverted to culverts or ditches or collected at a single location. Culvert or ditch length, gradient, cross section, and invert material may, however, tend to alter flow characteristics significantly. To a broader extent, a highway may inadvertently restrict a farmer's irrigation water, a city's potable water supply, or access to recreational activities on a lake. For reasons such as these, it is advisable to assure that all potential impacting water sources and potential impacted water sites have been spotted before attempting to locate (or relocate) a highway.

Particularly important for ecologically sensitive waters is the recommendation that highway corridors should be distant from potential receiving waters and ought to occupy a minimal amount of the area drained to a downstream waterway. Different land factors and terrain descriptions (as they relate to runoff), including the lack of water, may also become significant as design progresses. (Soil erosion, in somewhat of a paradox, can be extensive on desert roadsides after construction of a new or improved road.)

**FIGURE 6-1.** There are times when it is virtually impossible to construct a highway without disrupting a waterway. (Photo: Colorado Department of Highways.)

An architect must be forewarned concerning spots of potential difficulties, such as energy dissipation and varying flow velocities within each stream basin. Detritus, flotsam, and jetsam are often transported downstream and then deposited most inconveniently in places where the velocity and/or direction of the waterway changes. So, to be on the safe side, an architect should consider adopting a highway alignment that does minimal modification to the existing stream flows and drainage pattern.

## WATER ENVIRONMENT

Surface drainage and erosion control are two of the most important elements in highway architecture. But before turning our attention to these matters, a broader look at all aspects of the water environment through which a proposed or modified highway alignment will traverse is very prudent. We may encounter surface water, subsurface water, or even underground water. Understanding the quality of all these waters is relevant to how each is to be treated. Also, the quantities of each must be researched. So, prior to making drainage calculations, the overall hydrology has to be reconciled.

### Nearby Waterways

Where are (if any) the rivers (not to overlook those that are designated wild and scenic), streams, lakes, navigation channels, irrigation canals, and aqueducts? The survey maps should already have them accurately located. Not so exact would be the plotting of marshes, seeps, and other waterways that tend to have meandering bounds. Wherever possible, a highway location or widening should avoid major and problem waterways. Sometimes, this is not feasible.

### Underground Water

Springs, seeps, and variable-level water tables provide to highway maintenance personnel in many parts of the country their biggest stock of gripes. With this as a "given," it appears that attention to underground water ought not to be ignored during both preliminary and final design phases, particularly where a depressed roadway profile is being considered. Just how to cope with each potential underground water condition, once the pertinent details (depth, quantity, quality) have been ascertained, must be resolved as the design process continues.

### Armchairing

Sometimes it becomes advisable, when valid numbers are available, to note, directly on the preliminary water map/photo/sketch, the quantities of water flow expected at certain spots near the highway location. Gauging stations on rivers and streams may have collected data and accumulated historical runoff records over a considerable period. Later on, these data can be reduced to a program that will predict the frequency of floods likely to exceed expected values. Streamflow records may not be available in certain regions, or an on-site inspection of an existing waterway might not be feasible, especially for small channels with drainage areas of only several acres. The architect may then have to resort to good judgment (or consultation with an "old hand").

Every state and most local government agencies have departments, bureaus, or units that contain a great deal of background information, maps, and data relating to local water conditions. To stay ahead of the game, it's wise to bird-dog the following items (if applicable) and initiate the necessary consultation and coordination during preliminary design:

- Potable (culinary) water supply—Local health department and local water supply agency.
- Irrigation—County agent or regional agricultural bureau.
- Stream Modification—U.S. Army Corps of Engineers; also Coast Guard for navigable waterways.
- Wild and scenic rivers—Managing agency (i.e., USFS, BLM, National Park Service, or Fish and Wildlife Service) for compliance with Public Law 90-542, as amended.
- Wetlands—FHWA, *FHPM7-7-7*, and U.S. Army Corps of Engineers.
- Water quality—U.S. Environmental Protection Agency (EPA) and appropriate state government bureaus.
- Floodplains—FHWA, *FHPM 6-7-3-2*, and Federal Emergency Management Agency (FEMA).

### Additional Information

With water-related components laid out on a preliminary plan, the designer is better able to undertake the sundry tasks and operations involved in locating (or relocating) the highway. Where in place and available, a statewide or areawide computerized *Geographic Information System* (GIS) could be employed to relate waterways to other topographic elements. Also, the architect may wish to delve further into hydrology and hydraulics and to get a head start on drainage design by perusing *TR Record 1073* and FHWA *RD-82/063*.

## ECOSYSTEMS

Hydrology and hydraulics are not the limits of things to be considered. All water forms have an internal ecosystem and are part of a larger ecological environment.

68   Highways: An Architectural Approach

An overall appraisal must take this natural phenomenon into account throughout the design process (see Fig. 6-2).

During the preliminary design phase, the overall appraisal of water concerns is directed to alignment of the highway and how that alignment interacts with waterways. One example would be the condition that results when sudden surges of water have impacts on wetland systems.

## Wetlands

Wetland preservation continues to be a hot topic. The *Emergency Wetlands Resources Act of 1986* was intended to intensify cooperative efforts in managing wetlands. More recently, the U.S. Army Corps of Engineers, the U.S. Environmental Protection Agency, and the U.S. Fish and Wildlife Service have joined together in a National Wetland Symposium, which promises changes in existing wetland policies.

If wetlands are to become an important consideration in any specific design project, the designer may wish to consult *NCHRP Report 264*. To design a safe and efficient highway while at the same time attempting to protect wetlands, it is necessary to determine just what functions a specific wetland may perform (wildlife refuge, aquifier generator, aesthetic amenity) and what might be the effect of a highway on the wetland. Such a determination directs the architect to make the highway design compatible with its immediate water environment. A check and balance associated with this desired compatability between highways on the one hand and wetlands and coastal zones on the other is the ubiquitous "404 Permit" assigned to the U.S. Army Corps of Engineers. What this means is that any agency planning a wetlands impact by major construction activity must make an application to the U.S. Army Corps of Engineers for a permit (required by Section 404 of

**FIGURE 6-2.** A transportation facility adjacent to a waterway must take the ecosystem into account.

the Clean Water Act) to allow placement of material in a wetland during construction. The "404 Permit" is also involved with waterways having flows that exceed 5 cfs. There is an additional check on major construction in and near navigable waterways by the U.S. Coast Guard.

## Aquatics

Fish habitat requires a sufficient and continuous supply of nonpolluted water along with dissolved oxygen and a favorable range of water temperature. A balanced ecosystem plus resting pools (where the current is slow), riffles (where the current is fast), a clean spawning area, protective cover (so the fish can hide from predators), and an adequate food supply must be maintained in any freshwater stream diversion, relocation, or encroachment containing a significant fishery. Saltwater estuaries can be even more fragile and require additional cautions about disturbing habitat unique to the fish found therein.

Streamside vegetation is probably the most important factor in preserving a fish habitat, since it helps protect the banks from eroding, provides shade, contributes leaves, broken twigs, and other nutrients for invertebrates, and attracts insects that furnish fast food forage for finicky fish.

## WATERWAY DIVERSIONS

When it becomes necessary to entertain the possibility that a major waterway, or portion thereof, could be relocated to accommodate a transportation link, a good deal of care and attention must be directed to protecting the integrity of said waterway. Historically, too many ill-considered stream relocations were undertaken in conjunction with highway projects: stream relocatons that paid scant attention to changes in flow characteristics and that exhibited virtually no concern for ecological factors. Actions of this nature led to the many governmental controls and restrictions placed on highway designs that have the potential to impact waterways.

In the past—and into the present—many of those who have been technically trained looked upon stream relocation as an extension of highway construction and treated the matter accordingly. Frequently overdone in channel reconstruction was removal of riparian vegetation, an abasement that, among other things, permits transmission of more light to a waterway. Elevated summer temperatures result, to the detriment of salmoid fish and the invertebrates farther down the food chain. Architecture of stream relocations now takes this aspect, as well as a host of others, into account so that stream relocations become as extensively planned and constructed (albeit to different criteria) as any comparable edifice. Good stream relocation results in stable channels of adequate hydraulic capacity, a good ecological balance, restoration of stream banks, and appropriate aesthetic amenities.

One dilemma in architecturing a relocated, partially relocated, or just an impacted stream is the variability of stream flow. Flows in most waterways continually fluctuate; variations in magnitude between greatest and lowest flows of perennial streams can be as much as 100 to 1. This eventuality points out how it can become quite a challenge to attain the objectives cited in the previous paragraph.

The hydraulic elements to be addressed in a typical stream relocation can be found in a suitable text or handbook on hydraulic design. However, most such texts or handbooks do not address the important ecological and aesthetic elements. A nicely illustrated guide to "naturalizing" a channel reconstruction is *Restoration of Fish Habitat in Relocated Streams*, FHWA-IP-79-3.

## Preliminary Design of Relocated Stream Channels

It should not be necessary to point out that a relocated stream channel ought to be designed to carry about the same flow as the original natural channel. A rule of thumb places the design flow in the range from the 2-year (50-percent chance) flood, up to the 10-year (10-percent chance) flood. Greater flows would overflow onto the floodplain, being aware that impacting a floodplain is a risky business and should be approached by the architect armed both with trepidation and a copy of the appropriate highway agency procedures for floodplain encroachment.

Also, a relocated channel ought to approximate the hydraulic gradient of the original stream; natural streams are not always efficient as sluiceways. Natural channels wander and vary between steep and wide reaches. Widths change constantly, especially as channels first become filled with obstructions and then clear as a result of flooding or rechanneling. Uniform channels, at the other end of the spectrum, are easier to design and are appropriate for certain stream change applications. (In designing a uniform channel change for a stream, normal practice uses a trapezoidal cross section.) One adage that is always prudent for the highway designer: If it is not necessary to mess with a stream channel, don't. Where in doubt, the following conditions may warrant channel changes:

- The natural channel crosses the roadway at an extreme skew.
- Position of the natural channel endangers the highway embankment or adjacent property.
- The natural channel has woefully inadequate capacity, frequently overtopping.
- Retaining walls are impractical and the existing or proposed highway embankment encroaches on the channel.
- Channel widening in the vicinity of a highway (or at a highway crossing) to obtain material for the roadway embankment is coupled with the objective of improvement in the channel cross section or alignment.

Another distinction encountered is the perennial, as opposed to intermittent, stream. Since relocation of any stream channel will require environmental consideration, the perennial stream may not usually be disrupted during certain time periods. Construction is ordinarily suspended during the wet season of the year. If construction is necessary during such periods, sedimentation into streams may be minimized by installing debris basins, ponds, vegetative barrier strips, or silt fences. This aspect must be addressed in the construction specifications.

Two important items of information should be obtained at the beginning of the preliminary design of a relocated channel. They are the water surface elevation at the downstream end of the relocation, and the cross-sectional area and hydraulic radius at various points along the existing channel. These lead to the determination as to whether *critical depth* or *downstream control* governs. Water surface profiles for flow in nonuniform channels, such as natural watercourses or artificial channels with frequent changes of cross section and grade, cannot be computed by standard hydraulic methods. A channel having no appreciable length of constant cross section and grade allows no opportunity for conditions of uniform flow to be approached so that normal depth becomes a parameter of infrequent use. The critical depth, however, has increased relative importance. At this point, reference to an accepted hydraulic design manual or text is recommended. Additional references to consider include an analysis of stream channel design features contained in *TR Record 1127*, pages 50-60, a description of channel evolution in *TRR 1151*, pages 16-24, and the *Hydraulic Engineering Circular-20* put out by the FHWA.

## Use of the Computer

Computerization has reduced much of the previously laborious task of "trial and error" hydraulic design. The U.S. Army Corps of Engineers and other agencies have developed computer programs, each of which is intended to be a major computational aid for solving problems associated with a particular area of hydrologic engineering. The hydrologic elements must then be conditioned by environmental considerations, such as those discussed in *FHWA IP-79 3*.

## Scour and Erosion

An additional dictum that has already been stated: Do not change the length or gradient of an existing channel without a fantastically good reason. Replacing a long, sinuous natural channel by a shorter, straighter channel will increase the channel slope and usually decrease the channel roughness. Both of these changes tend to increase the velocity of the flowing water, sometimes sufficiently to cause excessive scour and head-cutting in the new channel. Scour and erosion are the two demons to be suppressed.

72  Highways: An Architectural Approach

Scour and erosion can be reduced through the use of properly designed linings. Linings may be rigid, such as portland cement or asphaltic concrete; or flexible, such as vegetation, geo-textiles, or rock riprap. Obviously, rigid linings are very seldom appropriate for natural channels. Flexible linings of erosion-resistant vegetation, properly selected geo-textiles, and rock riprap should be used when feasible. If vegetation is chosen as the permanent channel lining, it maybe established by seeding or sodding. Installation by seeding usually requires protection by one of a variety of temporary lining materials until the vegetation becomes established. Erosion along channel banks, in channel beds, and in areas where erosive tendencies exist, such as sharp changes in the channel direction, can be riprapped in several manners:

- Dumped stone (see Fig. 6-3);
- Hand-placed riprap;
- Wire-enclosed riprap;
- Grouted riprap;
- Concrete riprap in bags;
- Concrete slab riprap;
- Keyed stones;
- Compacted riprap; and
- Gabions.

The last three mentioned are probably the most effective for general use. *Keyed stones* are a fairly new construction technique that produces a tight uniform blanket of rock with a smooth surface. During the smoothing operation, the larger stones are fractured, thus producing smaller rock sizes to fill voids in the blanket. Smoothing or plating is carried out by a large steel plate (some 2 tons in weight),

**FIGURE 6-3.** Dumped stones are an economical means to control channel bank erosion—but they don't do much to preserve streamside vegetation.

which is used to compact the rock into a tight mass and to hone the revetment surface. Keyed stones are primarily used for protection of large stream channels. *Compacted riprap*, used for smaller channels and ditch protection, consists of durable, angular field or quarry stone generally no smaller than 4 inches and no greater than 8 inches in diameter. The stone is dumped into an excavated section that has been prepared and covered with filter cloth. The rock is manipulated to secure a regular surface of graded sizes and mass stability. Finally, the riprap is compacted by vibrating or pneumatic roller until there is a reasonably smooth surface with no protrusions. *Gabions* are compartmented rectangular containers made of wire mesh filled with stones. They provide assurance that the fill will remain evenly distributed after settlement. Gabions are used for overcovering stream channel erosion with drop structures. They become particularly useful in situations where wave action is anticipated. Geo-textile materials and applications change from week to week; however, *FHWA TS86-203* contains certain parameters for proper selection and methods of installation—specifically, Chapter 3: 3.5.5.5 *"Streambank Protection,"* 3.5.5.6 *"Wave Protection Revetments,"* and 3.5.5.7 *"Scour Protection."*

## CLOSURE

As stated at the beginning of this chapter, impacts created by water to a highway location, as well as by a highway location to water bodies, are two of the most significant impacts to be addressed during location and preliminary design. Final design will attend to drainage details, erosion control, and localized water impacts. They will be discussed in Chapter 11.

**References**
FHPM 6-7-3-2, *Location and Hydraulic Design of Encroachment on Floodplains*. Regularly updated. Washington, D.C.: Federal Highway Administration.
FHPM 7-7-7, *Mitigation of Environmental Impacts to Privately Owned Wetlands*. Regularly updated. Washington, D.C.: Federal Highway Administration.
FHWA. *Geotextile Engineering Manual*, Report TS-86-203. 1986. Washington, D.C.: Federal Highway Administration.
FHWA. *Hydraulic Engineering Circular-20*, Report IP-90-014. 1991. McLean, Virginia: Federal Highway Administration.
FHWA. *Local Design Storm*, Report RD-82/063. 1983. Washington, D.C.: Federal Highway Administration.
FHWA. *Restoration of Fish Habitat in Relocated Streams*, Report IP-79-3. 1979. Washington, D.C.: Federal Highway Administration.
Horner, Richard R., and Eugene P. Welch. 1982. *Impacts of Channel Reconstruction in the Pilchuk River*. Olympia: Washington State Department of Transportation.

Mitsch, William J., and James G. Gosselink. 1986. *Wetlands*. New York: Van Nostrand Reinhold Company.

NCHRP Report 264. 1983. *Guidelines for the Management of Highway Runoff on Wetlands*. Washington, D.C.: Transportation Research Board, National Research Council.

Transportation Research Record 1073. 1986. *Hydraulics and Hydrology*. Washington, D.C.: Transportation Research Board, National Research Council.

Transportation Research Record 1127. 1987. *Innovation, Winter Maintenance, and Roadside Management*. Washington, D.C.: Transportation Research Board, National Research Council.

Transportation Research Record 1151. 1987. *Hydraulic Erosion*. Washington, D.C.: Transportation Research Board, National Research Council.

# 7
# Pavements

If someone were to ask what the most essential component of a modern highway is, the answer would have to be the pavement. Pavements vary in their makeup because of the expectations placed upon them. A heavily traveled truck route would require significantly greater pavement strength and durability than a suburban cul-de-sac. This leads us to the matter of designing the pavement structure to satisfy the anticipated loadings and its expected life span.

Results of an extensive road test conducted from 1958 to 1960 in north-central Illinois by the (then) American Association of State Highway Officials (AASHO) provided the basis for nearly three decades of commonly accepted pavement design. It does not take a geology professor and a meteorologist to point out that these road test results reflected only one set of subgrade soils and only one climatic condition. Another limitation of the AASHO road test was that the effects of aging were not given adequate consideration. Constraints within the test have caused the findings to come under greater scrutiny as each decade passes. By 1984, when the Transportation Research Board (TRB) recommended a massive effort to stimulate faster implementation of highway technology, pavement design techniques had already become a major issue. The TRB recommendation led to the establishment of a cooperative endeavor by many sectors of the highway community—the *Strategic Highway Research Program* (SHRP, pronounced "sharp," not "shrup"). Originally, six priority areas were proposed, but in 1987, by combining some priorities, SHRP's areas of technical investigation were cut back to four. Two of the four priority areas deal with pavements; consequently, we can expect modifications to the long-accepted (occasionally dogmatic) beliefs and policies that have governed design and performance of highway pavements. Optimistically, with greater empha-

**FIGURE 7-1.** Shoulders, in addition to providing an area for vehicles to pull off the road, can help buttress the edge of the traveled way; absence of shoulders can lead to chipping away of the pavement edge.

sis placed on innovation (as opposed to tradition), highway pavements of the future should perform at a higher level and a lower cost.

## CLASSIFICATIONS

Pavements are generally divided into two major groupings, *unimproved* and *improved*, with subclassifications in each group.

Unimproved:

- Rutted wheelpaths;
- Graded;
- Graveled; and
- Oiled.

Improved:

- Flexible (asphalt and macadam);
- Rigid (portland cement concrete); and
- Modular (interlocking blocks and brick).

### Rutted Wheelpaths

Rarely would the highway architect be required to design a rutted one-lane road for ATVs, jeeps, or stagecoaches. Yet an already used path of this type may provide the initial alignment when an increase in use initiates a series of improvements to a particular route. Many of our major highways developed from such modest beginnings.

## Graded

In rural and some suburban locales, grading the terrain is all that is required to provide access for the limited number of vehicles expected. If fewer than about two dozen vehicles per day are expected to use a road, it isn't prudent to provide an all-weather wearing surface. Of course, there are exceptions. Should an influential county commissioner reside along a stretch of unimproved rural highway, financial resources are usually made available to provide an adequate hard surface to allow the commissioner's attendance at necessary governmental functions.

Grading of a very low-volume road is best accomplished by a competent and experienced dozer or patrol (grader) operator who appreciates the need to allow water to flow downhill without creating excessive gullies in the traveled way.

## Graveled

Usually, the next step to improve an existing graded road is to add gravel or other aggregate material, maybe even some calcium chloride, at critical soft spots. (Some locales have had success using enzymes to harden the surface of clay and gravel roads.) Often, the services of a highway designer can be enlisted to correct problems of grades that are too steep, inadequate sight distances, and poor drainage. A graded gravel road will often require pipe culverts, minimal delineation (posts, rock edging, occasional railing), plus a few directional and warning signs.

## Oiled

In some parts of the country, where soils are appropriate, the graveled road is provided with a three-oil treatment. After each application of oil, a light powdering of sand is added, followed by an interval of traffic before the next treatment. A semi-hard surface is the result—one that may require less maintenance effort than a gravel surface. Such a treatment can be satisfactory for the seldom-traveled rural road that passes by the residence of a former county supervisor or the local chairman of the minority political party. A three-oil treatment may not be expedient where the only aggregate sources are of particularly high-quality sharp-edged pieces. The sharp edges work to cut up the surface and tires of vehicles; rounded gravels from glacial, waterlaid, and windlaid sources generally work best.

## Upgrading

The next step upward has traditionally been to take a gravel- or oil-treated road and add a hard surface. As the desire to upgrade low-volume roads mounts while revenues decline, obtaining the most efficient use of available resources is paramount. Paradoxically, instances arise where placing a wearing surface on an existing road can actually cause detrimental effects. (One example frequently encountered: The subgrade becomes mushy as a result of the trapping of water because the road no longer "breathes" through what was formerly a pervious surface.) This means that there is an evident warrant to examine the existing base material on a gravel road before surfacing.

Reconstruction of a low-class road to a higher-type surface should, to obtain top value, consider use of existing in-place materials in spite of their possession of variations in quantity and quality. Cost-effective use of existing materials requires that their properties be evaluated to the appropriate degree.

Some other items to consider when upgrading a gravel road:

- Slopes—Avoid or contain landslides; clear bench sections; prevent further movement of embankments.
- Drainage—Establish roadside ditches; install culverts.
- Pavement geometrics—Establish uniformity in all components: thickness and width of pavement and shoulder materials.

The Kentucky pamphlet *When to Pave a Gravel Road* is a very fine guide to aid in determining warrants for upgrading an unimproved surface.

## FLEXIBLE

Flexible pavement surfaces have been associated (since 1870 in the United States) with bituminous materials. The generic term for dense pre-mix aggregates and bituminous materials is now equated with asphalt concrete. Asphalt concrete pavements are the most common in use today for all degrees of traffic, from light to very heavy. Most often they consist of several layers (or courses): wearing surface, base, sub-base atop a prepared subgrade (see Fig. 7-2). The move from flexibility toward rigidity is a desirable attribute in the evolution of design for heavy duty pavements. Yet, as pointed out above, traditional methods for designing flexible

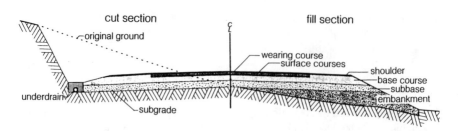

**FIGURE 7-2.** Asphalt pavements are built up in several courses; continual draining of the base course is essential.

pavement cross sections have been coming under critical scrutiny since the early 1980s, mainly because of the changes in the amplitude and mechanism in which truck loads are transmitted to pavements. Massive rutting on heavily traveled roads appears to be a recent phenomenon associated both with heavier truck loads and higher tire pressures.

**Rutting**

An indication of impending failure is considered to be distress. Rutted pavements usually fit into the distress category. To get a more comprehensive picture of just how truck loads and tire pressures affect pavements, the reader can refer to *TR Record 1227*. Two papers on heavy vehicles and three papers on tire pressures are contained therein. Asphalt pavements that, in the past, have been able to withstand the maximum allowable loads are now showing distress to an inordinate degree. Traditional high-type pavement design practice has been to construct a homogenous cross section from (often including) shoulder-to-shoulder, yet rutting only affects some 20 percent of this homogenous cross section. A number of research studies, including those under the SHRP aegis, are underway in an attempt to correct the condition and to prevent future occurrences. At the present time, maintenance forces are spending a considerable effort to combat and correct rutting on asphaltic concrete pavements—particularly those subject to heavy loads. If this issue can be addressed in the design stage, the maintenance effort may be reduced significantly. One of a number of value judgments suggests that, when designing flexible pavements (to be subjected to a great deal of semitrailer, doubles, and heavy truck traffic), increase the travel way width 6 to 8 feet. Thus, each year when the highway crews go out to paint the lane lines, they can advance them laterally 2 feet (to the right or left as the case may require). Although some rutting is likely to take place regardless, the maintenance forces will be better able to deal with a problem of only one-tenth or so of the magnitude that would be created by continual year-in, year-out tracking in the same ruts. After the second year, wheel path rutting proceeds rapidly. By limiting concentrated wheel path loads to one or two years, the pavement will, to some extent, rebound and provide better service even when the original wheel paths are again subjected to loading, four to six years later. Upgrading existing pavements and extending widths to move wheelpaths out of ruts is also gaining attention. Pages 7–13 of *NCHRP Synthesis 28* outline the essentials of design procedures for pavement widening less than one lane width; Appendix C of the same document shows various typical widening sections (some outdated) from six states.

Illegal overloading of trucks and trailers seems to be another contributing factor to rutting of flexible pavements. Recourse to the courts can be frustrating for a highway agency since, just to get a temporary injunction against a trucker who habitually runs illegal overloads, the agency must demonstrate that the trucker runs overloaded vehicles and that those vehicles cause "excessive" damage to highways under the jurisdiction of the agency. And, even if the trucker eventually has to pay a fine, it's just a "cost of doing business," since the fines are often much less than the additional revenues accruing from running illegal overloads. In addition, there are also "legal" overloads wherein a transporter can obtain a permit to run excessively heavy load(s) for special circumstances. (The highway, itself, is unable to distinguish between legal and illegal overloads.) This problem can be further exacerbated when, in some jurisdictions, the permits are issued by units outside of the highway agency.

In spite of many sanctimonious pronouncements by well-meaning transportation officials and law enforcement officers to the effect that increased trucking weights and illegal overloads will not be tolerated, the real world (where economics, followed by exertion of political muscle, plays a dominant role) says that trucks, tractors, trailers, and the loads they carry are more than likely to become of an even greater magnitude and concentration. Highway architects, builders, and maintainers are going to have to take this reality into account.

Rutting of asphalt pavements has become one of the biggest technical concerns in the highway community. Wheelpath depressions of more than one-half inch are considered ruts, and thus contribute to an overall pavement deficiency.

Broader specifications and design practices may permit certain other innovative actions or treatments to obviate the excessive maintenance effort now required to treat and correct rutting on major highways. Likewise, the traditional asphalt paving materials must also come under scrutiny. Some highway agencies believe that the solution is a better selection of aggregates and stiffer asphalt mixes. Other agencies favor the use of large stone mixes, in spite of resultant surface cracking that becomes more pronounced. Polymers and similar asphalt modifiers are used in certain instances, mostly experimental. At least one state highway agency has opted for inlaying the outside lanes of freeways (where heavy vehicles traditionally drive) with PCC pavement.

A prudent course of action would be for the architect to refer to the latest available literature when confronted with the design of a highway that is to be subjected to heavy, high-load traffic.

## RIGID

Including at least one course of portland cement concrete (PCC), thick enough to provide a high bending resistance, allows a pavement to be classified as rigid. Although PCC is high in compressive strength, its shear and tensile properties don't allow for extensive bragging rights. How and why some concrete pavements fail and others do not—under seemingly identical conditions—is still under conjecture. PCC

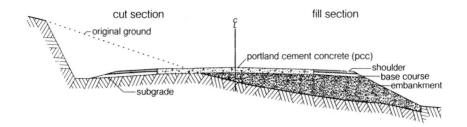

**FIGURE 7-3.** Portland cement concrete pavements are commonly used where heavy traffic weights are anticipated. (This simplified section is employed where there is a free-draining base.)

pavements have the highest value in situations where considerable heavy trucking is forecast. Consideration of PCC pavements need not be restricted to heavy loads and high traffic volumes, though. Contrary to views held in some circles, use of PCC in pavements for low-travel roads is a viable option. *TRR 875*, pages 21-26, describes successful experiences in Iowa on roads with ADT averaging 260 and paved with PCC.

Many roads have a PCC base and asphalt concrete wearing surface. Some were originally designed in this manner; others got that way when the original concrete surface showed symptoms relative to aging. When the PCC also serves as the wearing surface, it usually has the advantage of higher traction than most asphalt concrete pavements. Disadvantages include a generally rougher ride and higher noise level, although the latter is reflected on a greater scale within a vehicle than outside. Both reinforced and unreinforced portland cement concrete pavements have been used successfully.

Pre-stressed PCC pavements may be a thing of the future. *TR Record 1136* presents some findings regarding pre-stressed PCC pavements on pages 1-11. Pages 23-45 of the same publication contain two articles on another new concept: roller-compacted concrete pavement.

The initial cost for a rigid pavement is somewhat higher than for a comparable flexible pavement, but, over the first few years, the maintenance costs are low to nil.

## MAKING THE CHOICE

Even though political decisions often influence choices of construction materials where more than one product would provide satisfactory service, there are technical reasons for choosing which of several options to use when building a pavement. In dry climates, the compacting action of traffic tends to strengthen pavement sub-base, while the opposite occurs in wet climates. A climate with small daily and seasonal temperature changes favors the use of portland cement in concrete pavements. On the other hand, those regions having climates with large temperature changes produce excessive thermal expansion and contraction in portland cement concrete, which swings the pendulum in favor of asphalt concrete.

A long-term view ought to consider those resources available to maintain a satisfactory pavement. On roads surfaced with a bituminous cover, periodic sealing of this cover is required. Lower volume roads in many areas receive an application of a film of asphalt cement and a thin layer of small stone (often referred to as a bituminous or chip seal coat) to seal, bind, and protect the wearing surface. On higher volume roads, a thin resurfacing is usually added every few years. More information is contained later on in the chapter dealing with maintenance. One approach to determining optimum long-term investment in various pavement strategies is explained on pages 1-14 of *TR Record 572*.

## MODULARS AND SYNTHETICS

Cobblestones were once very common in city streets. Lately, the use of interlocking pavement blocks is coming into wider use as a surrogate. The Portland Cement Association has come up with design criteria and other information regarding their applicability.

Going one more step, a road can be improved by using materials heretofore not considered appropriate. Synthetic engineering fabrics or geo-textiles, as previously noted, have become increasingly important in civil engineering applications. Synthetic materials, as an integral part of a highway pavement, are generally used to provide separation between subgrade and base or sub-base and/or to provide a small measure of tensile reinforcement. Pumping action of traffic is alleviated to some extent as the fabric provides reinforcement, tensile resistance, and confinement of granular materials.

On minor rehabilitation projects, fabrics and membranes find use as an interlayer between deteriorated pavement and new asphalt concrete overlays. Reflective cracking is thus minimized.

## TO SUMMARIZE

To summarize the first half of the chapter:

- Gravel surfacing will usually suffice on roads where only 50 or so vehicles traverse in a 24-hour period.
- An oil treatment is probably appropriate for light (numbers and weight) traffic.
- A flexible pavement is best for most uses.
- Rigid pavement is the more popular approach where a good deal of heavy trucking is expected.

- Innovations in pavement selection and design should only be attempted with the endorsement of the sponsoring agency.

## BENEATH THE SURFACE

Without adequate support, no pavement can perform satisfactorily for long. The subgrade underlying all courses of a pavement structure must provide the requisite foundation.

### Subgrade

Uniform and non-altering strength throughout the lifetime of the roadway is the principal property to assign a subgrade. Too much to expect? Sometimes. Yet, with proper attention paid through design, construction, and maintenance, this goal can be approached, if not achieved. Soil, as we've already figured out, tends to be a substance of considerable variability. Inter-relationships of moisture, texture, gradation, and chemical makeup lead to a rather complex series of problems to be unraveled by the designer, the end product being optimum compaction of the subgrade soil. Sampling and testing are the standards used to assure that the subgrade can be (and is) compacted properly.

If the subgrade is of such character that it cannot be properly compacted, it may have to be partly replaced with different material or may need to be stabilized with bituminous materials, portland cement, selected granular borrow, membranes, or geo-textiles, or combinations thereof. The best place to start in determining suitability of subgrade materials is to investigate local practice: municipal and state specifications, test methods, and consultation with contractors and others familiar with pavement successes and failures in the area. Also, the AASHTO *Guide for Design of Pavement Structures* (AASHTO *Guide*) has some generalized procedures.

### Sub-base

The next level up from the bottom—applicable to high-type pavements mostly—is a layer of readily available materials (a cushion course of coarse consistency). In some localities there is little or no distinction between base and sub-base.

### Base Course

Constructed of bituminous-aggregate mix, gravel, stone, slag, soil-cement, old recycled pavement, or similar construction material, the base course is also an important component of the pavement structure—important not only to transmit loads from pavement surface to the subbase or subgrade, but equally essential to act as a purveyor to allow water to escape from underneath the wearing surface. Water buildup can cause serious consequences by changing soil characteristics of the subgrade and, in some areas, by leading to frost-heave.

## FROST

The AASHTO *Guide* has three maps that indicate the freeze-thaw-belt across the United States. On page II-72, of the six climatic regions showns, nos. III and VI are subject to freeze-thaw cycles. A similar belt (I-B, II-B, and III-B) appears in the climatic zones shown on page III-25. Then, on G-8, the map of North America frost and permafrost comes up with yet another, though similar, strip. Experiences over past years indicate that freeze-thaw cycles are prime culprits in pavement distress in an area slightly to the north of those belts shown in the AASHTO *Guide*. A rule of thumb suggests that freeze-thaw cycles are most prevalent in a belt roughly between the latitudes south of Interstate 80 and north of Interstate 20 (except in the Pacific rim states of California, Oregon, Washington, and Alaska). North of I-80 the roadbed remains frozen for 3–5 months of the year and only softens up during spring thaw; south of I-20 (and I-10) it rarely freezes. Therefore, the following applies mainly to the freeze-thaw belt:

- Frost action is one of the two principal causes of pavement failure (resulting in maximum seasonal distress during spring thaw).
- Frost heave is the raising of a portion of pavement as water beneath the portion turns to ice crystals and expands within the base and subgrade.
- Successive frost heaves during a winter of a dozen or so freeze-thaw cycles can mess up almost any pavement structure.
- There is an interaction of frost-heave and precipitation so that, by keeping the base and subgrade dry, frost-heave can be minimized.

*NCHRP Synthesis 26,* beginning on page 6, provides an analysis of the fundamentals of frost action. On page 32, five points of geometric design appropriate to the frost-heave belt are described.

Membranes, foam plastic, sand blankets, and certain geo-textiles have been used, sometimes successfully, to eliminate or moderate the conditions leading to frost-heave. There is also a bright side: good fortune reigns in much of the territory in the freeze-thaw belt because there is an abundance of free-draining glacial aggregates available for construction purposes.

### Permafrost

An entirely different ball game emerges when placing a roadway on ground that is frozen all year long. Thawing is to be avoided, and the soil must remain in its frozen state. Insulation of permafrost subgrade against any warming trend, including the conducting of heat from the sun's rays through a pavement section, is essential.

## DESIGN

The AASHTO *Guide* has been the basic tool for pavement design, but, since it covers an extreme range of conditions, it can become unwieldy for the architect or designer who has to come up with a pavement cross section for one specific condition. (Much like trudging through the Artic tundra in order to come up with the correctly patterned snowflake.) Therefore, if a designer chooses to apply the information contained in the AASHTO *Guide*, it is best to begin with Appendix B. Although both primary and secondary selection factors are cited in this appendix, the secondary factors probably exert more weight—especially in a sensitivity analysis—than those listed as primary factors. Appendix A details a glossary, and Appendix C is also worth perusing if the architect is interested in various alternative methods of pavement design.

Other usable references are also available. Some guidelines for design of continuous reinforced PCC pavement can be found on pages 14 and 15 of *NCHRP Synthesis 16*. A comparison of three approaches to structural design of emulsion-based flexible pavements is presented in *NCHRP Synthesis 30*, on pages 15-49. Designing pavements for anticipated overloads (both legal and illegal) is on the upswing, even though the widely accepted standard loading criterion remains the 18-kip (9-ton) axle. Pages 18-22 of *NCHRP Synthesis 131* outline, subjectively, how a designer might go about guesstimating some adjustments to account for future overloads. Updated design procedures for overlays are described in Hall, *et al.* (1992).

As inferred above, there has been increasing criticism of late directed at the concept that pavement design should be "rational," and thus, a science unto itself. This concept has led many highway agencies to continually beef up pavement thickness excessively so as to overcome any deficiencies in overload enforcement and subgrade consistency. Value-architecture principles would insist that, as a minimum, the latter deficiency—subgrade consistency—be zeroed in on for further examination. Inordinate water content in the subgrade is the most sensitive parameter. Thus, by free-draining or rapid draining of water beneath the wearing surface before it has a chance to permeate the subgrade, the cost of pavement structures can be reduced and the lifetime increased. Two excellent papers on this matter (if one is willing to overlook a somewhat sketchy economic comparison) are contained in *TR Record 1121*, pages 77-89.

One should be prepared for a lot of work in order to become expert in the field of pavement design. Many designers choose instead to opt for some available computer software to guide them through the process.

## REHABILITATION

Designing a pavement for a new alignment or completely replacing one on an existing alignment is being upstaged. Attempts to rehabilitate or reconstruct the existing pavement structure are now the center of attraction. Rehabilitation applies to both flexible and rigid types, although PCC pavements received the greater attention because flexible pavements are more amenable to continued maintenance procedures, as described later on in the chapter on maintenance.

When a PCC pavement has worn to the degree that it needs rehab, the standard procedure for many years has been to place a coat of blacktop over the surface. Initially, this procedure achieves the best of two worlds: continuing the rigid pavement concept while providing a smooth riding surface. But then, after several years, the cracks and joints in the PCC begin reflecting upwards and the smooth riding surface exists no longer.

To address this issue, a modified procedure "broken" into three mutually exclusive sub-procedures can be applied to the aging, distressed portland cement concrete pavement:

1. Crack-and-seat;
2. Break-and-seat; and
3. Rubblizing (Rubbleizing).

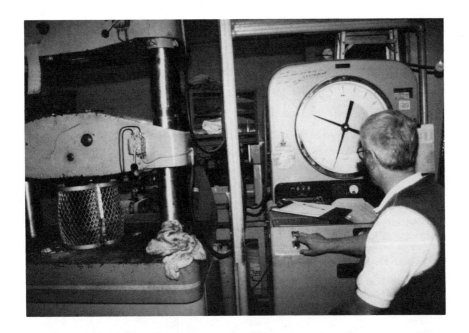

**FIGURE 7-4a, b, and c.** The traditional laboratory method for determining the strength of portland cement concrete is to subject a test cylinder (28 days after mixing) to an increasing load, recording the force necessary to break the cylinder.

Blacktop is then placed over the seated concrete, which acts as the base course. The final product is the replacement of a rigid pavement with a flexible one.

### Crack-and-Seat

For non-reinforced pavements, the PCC is cracked in place into small blocks (one to two feet across) and then "seated" with passes by one or more rollers. Three papers dealing with cracking-and-seating appear in *TR Record 1215*. A description of the procedure and some sample specifications may be found in *Thompson (1989)*.

### Break-and-Seat

A more vigorous procedure is required for mesh-reinforced PCC. The PCC is broken in place so that the mesh is sheared. As in crack-and-seat, the broken blocks are seated by rollers. Experience with break-and-seat in Kentucky is described by Sharpe, *et al.*, in *TR Record 1178*.

### Rubblizing

Where continuous reinforcing bars have been included in the old pavement, the PCC must be reduced to rubble. To accomplish this, a sonic hammer travels along and across the entire pavement, usually in longitudinal passes. Again, rollers are used to compact the rubble into an effective subgrade before the new asphalt surface is applied. A more complete description of the procedure appears in Flynn (1992).

### PCC Overlays

Asphaltic concrete overlays, as stated above, have been around for many years. More recently, the concept of using portland cement concrete for overlays has been gaining ground. Both bonded and unbonded concepts have been tried. A brief description of these procedures can be found on page 37 of *TR Record 1182*.

### Recycling

Keeping in tune with the times, the matter of recycling has become highly important in pavement rehabilitation. Generally, the goal of PCC recycling is to obtain suitable aggregate material for reuse in either new concrete or as a surrogate for selected base course material. Not only is some conservation of resources accomplished, but also the issue of solid waste disposal is addressed. *Yrjanson (1989)* provides a summary of PCC recycling. An added feature in the recycling effort is the use of flyash as an additive. Not only does this help dispose of a waste from coal-fired power plants, it also reduces alkali-aggregate reactions, which contribute to cracking in rigid pavements.

Recycling of asphalt pavements is directed at retrieving both aggregates and some of the bituminous materials in the older pavement surfaces. The U.S. Army Corps of Engineers *Recycling Primer* and the Wood, *et al.* paper in *TR Record 1178* can provide the highway architect with an entree into the asphalt recycling arena. From here, it's just a short step to the much broader field of environmental concerns.

## References

AASHTO *Guide for Design of Pavement Structures*. 1986. Washington, D.C.: American Association of State Highway and Transportation Officials.

*Asphalt Handbook*. 1965. Lexington, Kentucky: The Asphalt Institute.

*Design and Control of Concrete Mixtures*. 1979. Skokie, Illinois: Portland Cement Association.

El-Sheikh, Magdy, Joseph J. Sudol, and Rebecca S. McDaniel. 1990. Cracking and seating of concrete pavement on I-74. *Transportation Research Record 1268*:25-33.

FHWA *Geotextile Engineering Manual*, Report TS-86-203. 1986. Washington, D.C.: Federal Highway Administration.

Flynn, Larry. 1992. Contract drives Raleigh beltline rubblization. *Roads and Bridges* 30(1):36-44.

Fudaly, Thomas, Joe Massucco, and Tommy Beatty. 1990. *Large Stone Hot Mix Asphalt, State of the Practice*. Washington, D.C.: Federal Highway Administration.

Hall, Kathleen T., Michael I. Darter, and Robert P. Elliott. 1992. Revision of AASHTO Pavement Overlay Design Procedures. Paper read at 71st Annual Meeting of the Transportation Research Board, January 12-16, 1992, Washington, D.C.

Helping Hand Guide #2, *When to Pave a Gravel Road*. 1988. Lexington: Kentucky Transportation Center, University of Kentucky.

Janssen, Don. 1989. *PCC Mix Design* (WA-RD 176.1). Olympia: Washington State Department of Transpiration.

King, G. F. 1989. *NCHRP Report 324, Evaluation of Safety Roadside Rest Areas*. Washington, D.C.: Transportation Research Board, National Research Council.

NCHRP Synthesis 16. 1973. *Continuously Reinforced Concrete Pavement*. Washington, D.C.: Transportation Research Board, National Research Council.

NCHRP Synthesis 26. 1974. *Roadway Design in Seasonal Frost Areas*. Washington, D.C.: Transportation Research Board, National Research Council.

NCHRP Synthesis 28. 1975. *Partial-lane Pavement Widening*. Washington, D.C.: Transportation Research Board, National Research Council.

NCHRP Synthesis 30. 1975. *Bituminous emulsions for Highway Pavements*. Washington, D.C.: Transportation Research Board, National Research Council.

NCHRP Synthesis 131. 1987. *Effects of Permit and Illegal Overloads on Pavements*. Washington, D.C.: Transportation Research Board, National Research Council.

Peterson, Dale E. 1985. *NCHRP Synthesis 122, Life-cycle Cost Analysis of Pavements*. Washington, D.C.: Transpiration Research Board, National Research Council.

*Research Notes: Three Year Evaluation of I-40 Crack and Seat Experimental Project*. 1990. Phoenix: Arizona Department of Transportation.

Roman, R. J., Michael I. Darter, and Mark B. Snyder. 1985. *Procedures to Determine the Optimum Time to Restore Jointed Concrete Pavements*. Arlington Heights, Illinois: American Concrete Pavement Association.

Sharpe, Gary W., Mark Anderson, and Robert C. Deen. 1988. *Breaking and Seating of Rigid Pavements*. Transportation Research Record 1178:23–30.

Thompson, Marshall R. 1989. *NCHRP Synthesis 144, Breaking/Cracking and Seating Concrete Pavements*. Washington, D.C.: Transportation Research Board, National Research Council.

Transportation Research Record 572. 1976. *Pavement Design, Performance, and Rehabilitation*. Washington, D.C.: Transportation Research Board, National Research Council.

Transportation Research Record 875. 1982. *Design and Upgrading of Surfacing and Other Aspects of Low-Volume Roads*. Washington, D.C.: Transportation Research Board, National Research Council.

Transportation Research Record 1121. 1987. *Effects of Temperature and Water on Pavement Performance*. Washington, D.C.: Transportation Research Board, National Research Council.

Transportation Research Record 1136. 1987. *Pavement Design*. Washington, D.C.: Transportation Research Board, National Research Council.

Transportation Research Record 1182. 1988. *Concrete Pavements*. Washington, D.C.: Transportation Research Board, National Research Council.

Transportation Research Record 1215. 1989. *Pavement Management and Rehabilitation*. Washington, D.C.: Transportation Research Board, National Research Council.

Transportation Research Record 1227. 1989. *Rigid and Flexible Pavement Design and Analysis*. Washington, D.C.: Transportation Research Board, National Research Council.

TRB Special Report 225. 1990. *Truck Weight Limits, Issues and Options*. Washington, D.C.: Transportation Research Board, National Research Council.

Waterways Experiment Station. 1986. *Asphalt Pavement Recycling Primer*. U.S. Army Corps of Engineers.

Wood, Leonard E., Thomas D. White, and Thomas B. Nelson. 1988. *Current Practice of Cold In-Place Recycling of Asphalt Pavements*. Transportation Research Record 1178. pp. 31–37.

Yrjanson, William A. 1989. *NCHRP Synthesis 154, Recycling of Portland Cement Concrete Pavements*. Washington, D.C.: Transportation Research Board, National Research Council.

# 8
# Environment

One of the early stimulants to the environmental movement was the despoiling of nearby lands, waterways, and air by poorly conceived and improperly operated highways. Highways, en masse, became stigmatized. Once tarnished, it takes a long time and considerable good faith to get back in good graces. Highways have not accomplished this yet. In fact, highways, along with their traffic, are still considered some of the worst environmental polluters. Need it be this way? Can highways be designed and built to enhance the environment? Most folks would answer the first question with something uncertain, such as "maybe"; the second question would most likely get an emphatic "no way!" Let's look at what it might take to get fully affirmative answers to both questions.

## ENVIRONMENTAL ENHANCEMENT

Futurists have come up with a host of extrapolations relating to transportation. Some of the more prominent prognostications include greater vehicle efficiency, reduced emissions, electronic navigation, and maglev, all of which should contribute to less environmental degradation. Although there is no defined list of just what constitutes the total human and natural environment, the following topics are some of those usually found in the vanguard of environmental concerns.

### Aesthetics

This is the fun part. A prime reason for an architect to get involved with highways is to improve their looks. At the core is the alignment—how the road caresses the topogra-

**FIGURE 8-1.** A highway alignment that blends in with its surroundings is the first step toward attaining good aesthetics.

phy, how it is gently laid on the land—except for the city street and similar traffic arteries. Too bad. Most road and highway alignments have already been established: some good, a few excellent, but many that could be improved but probably won't be. So how about planting lots of shrubbery to hide the problems, not unlike that which the traditional architect does when the aesthetic appearance of his or her masterpiece edifice turns into a disaster? The linear nature of a highway makes this tack highly impractical. Most of the time, a few spot improvements are all that is needed to restore a bad alignment to at least an acceptable level.

After the alignment is corrected, if it needs to be at all, the pavement, shoulders, and roadside should be tended to. Although there are many constraints imposed by safety considerations, capacity requirements, and available financing, the architect has an Olympian opportunity to be innovative, imaginative, and creative—traits that highway engineers rarely have, or, if they have, rarely use.

One aesthetic aspect that seems to come under a lot of flack from highway technologists is the proximity of trees to the traveled way. This flack is often well founded. Trees too close to the pavement can be real hazards. In addition, the shade provided can provoke danger in two ways: intermittent shade may cause temporary blinding to drivers trying to make out the road ahead, and prolonged icing can be a problem in well-shaded tree shelterbelts.

**Air Quality**

Something approaching half the ozone precursors and three-quarters of the carbon monoxide emissions in larger urban areas come from motor vehicles. How in the name of all things sane can highway construction and the increased traffic that goes with it improve air quality? Certainly not directly. Indirectly, the trend towards alternative fuels, particularly methanol, can lead to cleaner air while concurrently reducing agricultural waste and some of the big city garbage. This will all come about when it becomes economically advantageous to do so; the technology is presently available.

## Archeology and Paleontology

There have been documented cases from most parts of the country in which excavations for highway construction have uncovered fossils and evidences of mankind's early settlements. These may never have been discovered had it not been for the excavations.

## Endangered Species

As a result of the environmental documentation required in connection with major highway construction, several endangered animal species and a considerable number of rare plant species have been identified and provided with protection.

## Farmlands

Prime and unique farmlands are not to be removed by new highway locations without extenuating circumstances. However, their productive use is highly dependent on farm-to-market roads.

## Floodplains

There have been occasions where highway alignments have been placed purposely to isolate or contain floodplains. Other cases include elevating highways above the floodplains in order to provide transportation and connections to isolated communities during times of high water.

## Historic

Many historic edifices and districts have become run down over the course of time. Occasionally, the intrusion of a new or improved highway presents the opportunity to spruce up these historic fixtures and add to their longevity.

## Noise

It is extremely far fetched to contend that highway traffic improves the sonic environment. Even the attenuation measures now in vogue are not always popular. Let's drop this topic for now.

## Waste Management

Disposing of and recycling waste products are two particulars in which highways have more recently become involved. Because of the enormity of some roadway embankments, they have been used as repositories for certain waste materials. By distributing the materials throughout the embankments, the concentration is reduced to a level generally considered safe. Although rare, this approach can be appropriate under specific circumstances where it is possible to avoid contaminating the groundwater. An example can be found in Roshek (1992).

Recycled products such as flyash and nondegradable rubber products are commonly incorporated—the former in concrete and backfills, the latter in erosion control, shoulder reinforcement, pavements, and retaining wall tie-backs.

## Water Quality

Here is another instance where a highway alignment can contribute in a positive way to cleaning up the environment. Much of the time, industrial wastewater and agricultural effluent is allowed to flow across the land and into the soil, eventually winding up in the groundwater. Highway alignments running transverse to these flows will intercept the polluted waters and channel them into drains and culverts where, with a little concern and foresight, the pollutants can be segregated, treated, or otherwise removed from endangering the environment. So far, very little has been done to accomplish this goal, but the opportunity is there. When this possibility becomes more widely known and accepted, many existing roadway prisms are going to be retrofitted to enhance water quality. There is even the possibility that, with proper monitoring, certain replacement wetlands can be regularly recharged to provide wetlands of a higher quality than the ones being replaced.

Modifying stream channels in an environmentally sensitive manner has already been described in Chapter 6.

## Weighing the Options

If the above examples seem rather strained and forced, it's because they *are* strained and forced. It is not to imply that highways must be a continual drain on the environment, but rather to plant a seed in those who will be shepherding highway projects of the future that environmental degradation does not have to go with the territory.

As a further point, if possible, pollution should be attacked at the source, before pollution is created. Traditionally, technology has concentrated on catching pollutants where they are discharged into the environment. By using our resources more efficiently and by ensuring that one means of pollution prevention does not create another pollution elsewhere, the highway community may some day be able to categorically state that highways are good for the environment.

## "IT'S THE LAW!"

Virtually all environmental issues have achieved legal status; surprisingly, some have been subject to federal and state laws for many decades. The following capsule contains but a miniscule listing of the various attempts to legislate environment; it does contain most of those that have had an impact on highways.

1899—*Rivers and Harbors Act*, which was designed to control dumping and dredging in navigable waters.
1928—Congressional authorization of what was to become the George Washington Memorial Parkway from Arlington to Mt. Vernon.
1934—*Conservation of Wildlife, Fish, and Game Act*.
1934—Natchez Trace Parkway through parts of Mississippi, Alabama, and Tennessee was given the go-ahead by Congress.
1958—*Fish and Wildlife Coordination Act*.
1965—*Federal Highway Beautification Act*, which began controlling billboards near highways.
(Also, about this time, Lady Bird Johnson began lobbying for highway beautification.)
1966—*National Historic Preservation Act*.
1966—*Transportation Act of 1966*, Section 4(f) of which required protection be given to park and recreation areas.
1970—*National Environmental Policy Act of 1969*. (Requirements instituted for environmental impact statements and establishment of the three-person Council on Environmental Quality (CEQ) to advise the president on environmental matters.)
1970—*Clean Air Act*.
1972—*Coastal Zone Management Act*.
1972—*Federal Water Pollution Control Act*.
1973—The Environmental Protection Agency promulgates EPA 430/9-73-007: *Process, Procedures, and Methods to Control Pollution Resulting From All Construction Activity*.
1973—*Endangered Species Act*.
1973—*Moss-Bennett Act*, to preserve archeological sites and historic properties.
1974—*Safe Drinking Water Act*.
1974—Establishment of the Advisory Council on Historic Preservation.
1976—*Resource Conservation and Recovery Act*, directed at solid waste disposal.
1977—Presidential Executive Orders 11988 and 11990, concerning floodplains and protection of wetlands, respectively.

1977–*Clean Water Act*, which required preparation of "404" permits. (Amendments to the *Clean Water Act* are contained in the *Water Quality Act of 1987.*)
1977–*Amendments to the Clean Air Act.* (Additional amendments came in 1990.)
1978–*Amendments to the Endangered Species Act.*
1979–U.S. Fish and Wildlife Service Consultation Procedures promulgated.
1979–Council on Environmental Quality establishes regulations (to replace guidelines) for preparing environmental documents.

It was the Council on Environmental Quality regulations of 1979 that put some teeth into environmental documentation. From 1970 to 1979, environmental paperwork was cajoled by guidelines (actually suggestions) from the CEQ. It was during this span that most federal agencies, in the interest of pursuing their respective charges, began to abuse the spirit of the CEQ guidelines. Because the federal aid highway program was primarily carried out by state-level agencies, the abuses were multiplied fiftyfold. Effectively, the highway community had shot itself in the foot by mishandling too many environmental issues in the name of getting more pavement down. The CEQ regs of 1979 not only put a damper on overzealous road builders, they also spawned some enormous headaches for highway agencies and brought the EPA directly into the loop by making them the repository for environmental impact statements. The paperwork, coordination with other federal and state agencies, mandated review periods, and assorted bureaucratic hangups made environmental processing the lengthiest of all the procedures necessary to get a major highway project from initial concept to final completion.

The *Environmental Flow Charts* prepared by the Federal Highway Administration can give an idea of the paperwork jungle one has to go through to satisfy most federal environmental requirements.

## Documents and Coordination

Not all non-highway federal and state agencies are enthused about spending taxpayer money on highway construction. Most of these bureaucracies have been empowered by law and regulation to throw their weight around whenever any major transportation facility is proposed in or near their dominion. Usually their input is valuable and constructive, but sometimes they will delay or disfigure a worthwhile highway improvement to the benefit of no one. This brings up the poser as to just how involved the highway architect should become in the environmental paper pushing game: not at all, moderately, or completely. The last-mentioned is the only course to follow in order to protect the highway concept from being nibbled away or thoroughly decimated by excessive bureaucracy.

Most major environmental documentation efforts–environmental impact statement (EIS) or environmental assessment (EA)–are put together by specialists, either within the highway agency itself or by outside consultants. In too many cases, the highway planning and design staffs are involved only in a cursory manner.

A bad scene.

They should be involved up to their armpits as a minimum. Two basic reasons are presented for this assertion:

1. Input to the environmental process from the highway specialists is vital to maintain the rationale and concept for a worthy highway proposal.
2. A knowledge of what transpired during the environmental process and how everything fits into the overall picture is essential to all of the decisions made on a day-to-day basis throughout design and construction.

Thus we've come to other substantial reasons for an architectural, rather than an engineering, approach to highway project development.

## MITIGATION MEASURES

Paying a return visit to some of the topics discussed further back in this chapter, there are some basic measures that a highway architect should see taken in order to permit the assigned highway project to fit well into the human and natural environment, whether or not these measures are mandated by law or regulation.

**Visual Resources**

Although several methodologies have been proffered to identify the significant visual aspects, there doesn't yet seem to be a compendium of what, how far, and how much to incorporate into a typical highway endeavor.

A view *of* the road tends to be static. True, there is some vehicular motion observed, but the road alignment and structure remains in place. The view *from* the road, conversely, is full of motion and change. The difference between "of" and "from" is quickly likened to the difference between still photos and movies, or between the interior of an art museum and the exterior of a protest rally. That is to say, they must each be treated differently.

Highway bridges, retaining walls, and rock facings are three prime elements to enhance the view of the road in rural settings (see Fig. 8-2). (We've already stated the importance of alignment, both horizontal and vertical.) Urban and suburban settings, on the other hand, rely more on color, immediate roadside treatments, and lack of clutter (see Fig. 8-3). All of these elements can be done aesthetically and, at the same time, made to be compatable with necessary safety and capacity constraints.

Difficulties arise when efforts are directed towards enhancing the view from the

98    Highways: An Architectural Approach

**FIGURE 8-2.** Wood and stone go well with a rural setting. (Photo: Virginia Department of Transportation.)

road. The observer is in motion (except maybe in Los Angeles during commuter hours). Vehicular passengers may enjoy extended sight-seeing, but the driver ought not to be diverted from that task of watching the road ahead (and behind) for a length of time any more than an instant. This directs the aesthetic approach to one of providing, where feasible, the best viewing directly ahead and devoid of surprises. It's all right to approach a magnificent vista or wide panorama so that it does not distract the driver from the tasks of watching for other vehicles and observing the roadway geometrics along with warning signs. It may not be all right to open those same vistas or panoramas in such a dramatic manner as to cause the vehicle operator a sudden spectacular rapture with the magnificence of the scene ahead.

Highways are dangerous and hazardous. Aesthetic amenities must take this

**FIGURE 8-3.** Roadside treatment and absence of clutter are assets in an urban setting.

consideration into account at all times. Some thought could be given to the effect of motion, repetition of similar elements seen from the road, length of observation period, frequency of travel through a specific location (commuter vis-a-vis tourist), background, color variances, dimension or scale, traffic volume, characteristics of the majority of vehicle operators, and position of the sun during various times of day. This art form is still in its infancy, so the architect has a clean white pad of paper (or blank computer terminal screen) on which to begin.

## Ecology

Vegetation and wildlife within and immediately adjacent to the highway corridor are often overlooked by the engineers who direct their primary attention to earthwork, pavements, and drainage, items that are certainly important, but not the beginning and end of all the concerns that a highway generates. One example that is becoming of increasing interest is the urban heat island effect.

Huh?

Urban heat island effect is the increase in urban area temperatures caused by replacing vegetation with paving, roofs, and other heat-absorbing surfaces. It is contended that since 1940 there has been an attributable downtown warming of up to 1 degree farenheit per decade. It is further contended that from 5 to 10 percent of the urban electric demand in summertime is to cool buildings to compensate for the heat island phenomenon. Also, it is contended that the heat island effect contributes to an increase in smog. All these contentions may well be exaggerated, but the effect is, nevertheless, real. Planting vegetation, where appropriate and not hazardous to traffic, within highway and road right-of-ways is being considered as one means of countering the heat island effect.

Other examples of ecological enhancement opportunities abound, but they are constrained by the economic climate. Cost-effectiveness of roadside vegetation needs to be established and documented before serious consideration is given. This is not an easy task, particularly when confronted with a political and technological mindset that says that available highway monies should be spent on more asphalt and PCC. One means of countering such a mindset may be to illustrate effectively that roadside vegetation fills physical and psychological needs to both motorists and the community. (Run it up the flagpole; there may be enough empathy in the community to keep the flagpole from being chopped down.) Whatever the decision on what and how much is to be provided, roadside plantings ought to be considered a managed resource rather than simply controlled vegetation.

Federal and state governments are pushing for a greater degree of ecological awareness in environmental design. AASHTO, TRB, and the FHWA have all promulgated manuals and guides for accomplishing this end. More about ecological aspects is presented further on in Chapter 14.

100    Highways: An Architectural Approach

## The Other Senses

Much emphasis has been placed on sight. Nevertheless, we should not ignore hearing and smell when we analyze the aesthetic impact of highways.

Lowering noise levels is difficult, expensive, and not always successful. Earth berms and noise walls are the most commonly used highway noise suppressants, but the walls tend to reflect noise so that some receptors, both within and outside the right-of-way, are given an added dose of decibels. Mitigation tests on noise walls have demonstrated that canting the walls a slight bit from vertical and placing sound-absorptive materials on the wall faces more effectively reduces some of the noise impact to adjacent properties. Transparent noise barriers of glass or plastic have been tried in Canada and Maryland, but as of this writing have not generated much of a bandwagon. Additional discussion on suppressing traffic noise is presented in Chapter 14.

An aspect of noise propagation that usually gets overlooked is the atmospheric temperature inversion effect. This is why morning commuter traffic is far more intrusive than in the afternoon when noise refracts upward. Other weather factors, such as wind and turbulence, contribute to the complexities of noise generation.

Pollutants from vehicular travel that produce unpleasant smells are the hydrocarbons. Trucks, buses, and other diesel-powered vehicles contribute the higher share of hydrocarbons. The architect should also realize that construction-induced odors, such as those produced during bituminous paving operations and excavation of certain soils, can be addressed in the construction specifications. As an example, permitting a contractor to locate an asphalt batch plant next to the municipal rose gardens would not be a prudent move, regardless of accessibility to the project and availability of the site.

**FIGURE 8-4.** Noise walls can provide both acoustical and visual suppression. (Photo: Contech Construction Products, Inc.)

## INTERDISCIPLINE OR MULTIDISIPLINE

Some regard these two terms as synonymous; others contend that there is enough of a variance to warrant some differentiation between the two. So, to offer a compromise, we'll presume the following is a satisfactory differentiation:

- Interdiscipline—Members of diverse disciplines interacting directly with each other in developing solutions.
- Multidiscipline—Members of diverse disciplines providing solutions restricted to their own expertise, without necessarily involving each other.

For the long term, an interdiscipline approach to environmental issues usually works best, the rationale being that representatives from each diverse point of view have been given the opportunity to understand where the other points of view are coming from. The multidiscipline track should be reserved for instances when a quick fix is needed to get a project designed in a very short time frame.

## ENVIRONMENTAL FULFILLMENT

Although most of the environmental documentation takes place during the planning and preliminary design phases, the actual consideration of all environmental issues is a continuous undertaking, ranging all the way through construction and operation of the facility. This is another reason for providing continuity in project management and for having a highway architect shepherd a project through its many phases of development. The reader will note, therefore, that environmental issues keep cropping up in every chapter.

**References**

AASHTO. *A Guide for Transportation Landscape and Environmental Design.* 1991. Washington, D.C.: American Association of State Highway and Transportation Officials.

FHWA. *Environmental Flow Charts*, Report IP-87-09. 1987. Washington, D.C.: Federal Highway Administration.

FHWA. *Guidance for Preparing and Processing Environmental and Section 4(f) Documents,* Technical Advisory T6640.8A. Washington, D.C.: Federal Highway Administration.

FHWA. *Visual Impact Assessment for Highway Projects*, DOT-FH-11-9694. 1981. Washington, D.C.: Federal Highway Administration.

Fire, Frank L., Nancy K. Grant, and David H. Hoover. 1990. *SARA Title III, Intent and Implementation of Hazardous Materials Regulations.* New York: Van Nostrand Reinhold.

Henderson, Jim E., Richard C. Smardon, and James F. Palmer. 1988. U.S. Army Corps of Engineers Visual Resources Assessment Procedure. *Transportation Research Record 1189*: 67–71.

NCHRP Legal Research Digest 15, *The Application of NEPA to Federal Highway Projects*. 1990. Washington, D.C.: Transportation Research Board, National Research Council.

NCHRP Synthesis 87, *Highway Noise Barriers*. 1981. Washington, D.C.: Transportation Research Board, National Research Council.

Rieley, William D., *et al.* 1989. Road Planning and Design. *Landscape Architecture* 79(3):56–61.

Roshek, Michael W., and Scott Goodwin. 1992. Recycling Project: Concrete Grinding Residue. Paper read at the 71st Annual Meeting of the Transportation Research Board, January 12–16, 1992, Washington, D.C.

Transportation Research Record 1166. 1988. *Issues in Environmental Analysis*. Washington, D.C.: Transportation Research Board, National Research Council.

Transportation Research Record 1176. 1988. *Research on Noise and Environmental Issues*. Washington, D.C.: Transportation Research Board, National Research Council.

Transportation Research Record 1245. 1989. *Transportation of Hazardous Materials*. Washington, D.C.: Transportation Research Board, National Research Council.

Transportation Research Record 1255. 1990. *Energy and Environment 1990: Transportation-Induced Noise and Air Pollution*. Washington, D.C.: Transportation Research Board, National Research Council.

Zeigler, A.J., *et al.* 1986. *Guide to Management of Roadside Trees*, FHWA-IP-86-17. McLean, Virginia: Federal Highway Administration.

# 9
# Geometry

"Let it lie lightly on the land."

Highway architects may be expected to observe such a philosophy when realigning an existing road and, in rare opportunities, laying out an alignment for extended lengths of new highway. Because there is so much interdependence between geometrics and all the other features that go into the makeup of a highway, a two-step procedure in developing the geometry usually works best. Step one is referred to (in many highway agencies) as *location*. Implied in the term is a preliminary cut at geometrics. Much like a draft of an important written document, preliminary geometry often goes through many small changes before the final "document" is published. Final geometry gets "published" near the end of the design phase. Sometimes final geometry, or portions thereof, gets set in the field during construction. Even maintenance activities have been known to alter some roadway geometries.

## PRELIMINARY LINE AND GRADE CONSIDERATIONS

Aesthetics are but one component in placing an alignment; they do play a prominent role in preliminary geometry. However, the highway architect must also be aware of other components in order to make judgments on priorities and trade-offs. Our first priority must be the highway users.

## The Driver and the Vehicle

Paramount concern when developing (or redeveloping) highway geometry must be directed toward the individuals who are most directly affected by design decisions: the drivers of the vehicles that are to use the highway. As an individual drives along highways designed to be commensurate with his or her skills (and constraints), operation can be complementary, but, where a design is not in harmony with a driver's normal abilities—and here we must consider the driver near the lower limit of normal abilities—opportunities for problems increase. Hazardous or, at best, inefficient operations are the consequences where a highway's geometry outranks the capabilities of vehicle operators.

Coupled with driver limitations, the capabilities of the various classes of vehicles that will use the highway deserves strong consideration. Thus, acceleration, braking, and steering or cornering characteristics of vehicles have long been judged to be important parameters for determining design criteria. What class of vehicles should be used to establish those design criteria?

Trucks and semis.

Superior acceleration and braking rates, as well as better cornering characteristics, deem that passenger cars rarely exert control over design features.

Reconstructing or rebuilding an existing highway doesn't obviate the desirability of improving or updating the vertical and horizontal geometries. Sometimes a complete redo is in order; at other times the insertion of short segments of new geometry would be adequate. Safety hazards are a prime reason for improving geometrics. Poor geometric elements show up in three of the five most-often-cited roadway safety hazards. Many older alignments contain some built-in hazards that were not hazards when passenger vehicles were built higher and operating speeds were set lower.

## PLOTTING THE ALIGNMENT

At this stage, the overall topographic considerations (water-related items in particular) have already been drawn up on suitable maps or drawings as a plan view; the preliminary centerline of the road can now be superimposed on the plan view as an intermittent sequence of circular curves connected by tangents. Hopefully, the maps and/or topographic drawings are of a size, scale, and configuration that display continuous stretches of the proposed roadway. Using the available topographic information, the architect makes adjustments in both horizontal and vertical alignment jointly as work on preliminary geometry is done and redone graphically (CADD when available and cost-effective).

AASHTO-90 *Green Book* is the most widely accepted resource on highway geometry. Nevertheless, as with most in-depth reference works, there is a great deal of information to be sorted out, and no single highway design would find applicability in all of its 1,044 pages. In the first place, AASHTO-90 *Green Book* separates the basic design parameters for the various road hierarchies. This means that a little-traveled rural road would not be subject to the same yardsticks as would a highly traveled urban expressway.

**Basic Parameter: Design Speed**

Some sort of primal underpinning is obligatory to set up a frame of reference for permissible minimums and maximums. *Design speed* is that basic parameter. Continuous operation at a presumed vehicle speed under expected roadway conditions determines the design speed to be used as the overall control. In this way, design speed establishes limiting values for sight distance, curvature, maximum grades, width, and clearances. However, it must be remembered that many subjective elements come into play. Driving at high speeds in locations that might be unfamiliar to a driver; a vehicle operator who may be under adverse environmental and emotional conditions; an undemanding, monotonous highway leading a driver to inattentive operation—all are factors to be dealt with. They imply that minimums and maximums dictated by design speeds are not to be thought of as rigid straightjackets, but rather as guidance to aid in formulating geometric design. Design speed criteria have been fairly well established for the various road hierarchies and should be treated as "givens."

**Alignment Coordination**

Many beginning designers make the mistake of plotting a nice looking horizontal alignment on the monitor screen (or the working drawings) and then trying—with limited success—to impose a vertical profile to fit the conditions as best they can. Horizontal and vertical geometries are best not designed independently. Since they must be complementary, emergence of uncoordinated horizontal/vertical geometry can ruin the best parts and accentuate the weak points of each component. Alignments (horizontal and vertical) are extremely important in the permanent design of a highway. Concurrent development of horizontal and vertical alignment usually reduces the cost of constructing a highway. Therefore, it is best to keep in mind that horizontal alignment must be coordinated with the proposed grade line.

Topography affects horizontal positioning to some extent, but is more apparent in its impact on roadway profiles. Moreover, interrelationships between horizontal and vertical determinations continually come into play. For instance, sharp horizontal curvature had better not be introduced near the low point of a sag vertical curve because vehicular speeds, mainly of trucks, are highest at the bottom of downgrades: a dangerous situation, especially at night, because of the illusion of a foreshortened road ahead. For slightly different reasons sharp horizontal curvature should not be placed near the top of a crest vertical curve. Such placement contains the danger that the driver may be unable to view the curve, especially when beams from the headlights go straight off into the night sky. If in doubt, begin a horizontal curve before and end it after a vertical curve. Another good rule of thumb is to maintain consistency and avoid sudden departures from prevailing geometry. Consistent alignment, in which both horizontal and vertical components satisfy limiting criteria, has to be considered basic.

**Horizontal Considerations**

Although both horizontal and vertical geometry should be developed simultaneously, there are some aspects that pertain only to the horizontal component. For openers, an architect should begin by setting and locating the curves rather than the tangents. Neither superelevations (banking on curves) nor side friction should exceed specified safe values. This means that there are limits on degree of curvature that may be used in connection with any designated design speed. Further, avoiding sudden reversals in alignment makes it easier for a driver to keep within his or her own lane; reverse curves cannot be superelevated properly. A modest reversal in alignment may be had by inserting a short tangent between reverse curves (see Fig. 9-2).

Reiterating a point already emphasized, geometry ought to be consistent. Abrupt changes from flat curves and long tangents to sharp curves can create high hazard sites. Compound curves having radically different degrees of curvature or its stablemate, the "broken back" curve (containing a short tangent between two curves in the same direction), is not good geometry. Most drivers do not feel comfortable with consecutive curves in the same direction. Successive curves in opposite directions (separated, of course, by significant tangents) on the other hand reinforce the subliminal desire of drivers to follow a rocking back-and-forth pattern. Avoiding "broken back" curves is a holdover from railroad practices dating from the nineteenth century. In the days when passenger travel by rail was an acceptable mode, it was found that on long trips, passenger fatigue set in on lines that had many "broken-back" curves. It seems that the human body adapts much better to the undulating rhythm of rolling from side-to-side, which is the case when railroad cars rock from one side to horizontal to the other side. However, the "broken-back" curves caused railroad cars to rock to one side, return to vertical and then rock back to the same side: most disconcerting and fatiguing when it went on for hour after hour. A similar rationale follows for highway travel (provided that the soothing undulations don't put the driver to sleep).

In highways with higher design speeds, centrifugal forces generated on flat curves can be compensated by a modest superelevation. But should a normal cross slope of 2 percent or so be retained, a dangerous reverse superelevation occurs on

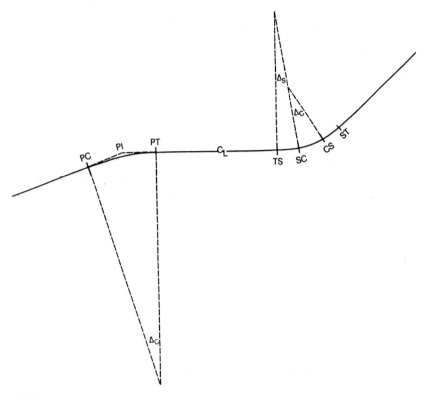

**FIGURE 9-1.** Common nomenclature for horizontal curves:
  D = Degree of curve = 5729.58/ radius in feet;
  LC = Length of curve;
  PC = Point of curvature;
  PI = Point of intersection of tangents;
  PT = Point of tangency.
For spiraled curves (arc definition):
  LS = Length of spiral;
  TS = Tangent to spiral;
  SC = Spiral to curve;
  CS = Curve to spiral
  ST = Spiral to tangent.

The highway control line (CL) is usually the centerline of pavement on two-lane roads or the edge of the lane adjacent to the median for divided highways. Convention directs that stationing (numbering every 100 ft.) proceed west-to-east or south-to-north.

curves to the left. This raises the question of whether or not to provide a reverse crown or to bank relatively flat curves, a dilemma that is usually resolved later when the geometrics are finalized.

108   Highways: An Architectural Approach

**FIGURE 9-2.** Reverse curves should be separated by a short length of tangent to allow for the necessary change in superelevation. (Photo: Virginia Department of Transportation.)

## Vertical Considerations

Tangent lines and two varieties of parabolic curves are put together to make up a highway profile. Sag vertical curve limits differ from those for crest vertical curves in more than just the way the bight faces. Thus, design values for sag vertical curves are not the same as those for crest curves. Headlight sight distance has become the selected criterion for establishing controls for vertical curves. Headlights on most passenger vehicles are mounted some 2 feet above the pavement, but below height of the driver's eye, a condition that results in sight distances over crests being shorter at night than in daytime. (Here then is a condition where passenger car characteristics override those for trucks and semis, which have higher headlight placement.) We can expect that minimum lengths for crest vertical curves would vary with design speed. Longer curve lengths are preferable because, in cases where grade changes over crests are too sharp, changes in gravity forces cause discomfort to vehicle operators and passengers.

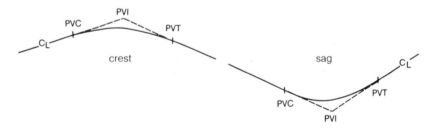

**FIGURE 9-3.** Common nomenclature for vertical curves: PVC = Point of vertical curvature; PVI = Point of vertical intersection, PVT = Point of vertical tangent.

## Rectifications

To place a highway lightly on the land isn't always easy. Many of the nation's highways have not been laid out as thoughtfully as they could have been. Corrections have been, are being, and will be made to poor geometrics. But first, there must be a determination as to what constitutes less-than-good roadway alignment. One example of mediocre geometrics may be seen on the accompanying annotated photograph of a freeway segment in rolling country (see Fig. 9-4).

Here we have a demonstration of what can happen when there is minimal coordination between horizontal and vertical geometry. Evidently, the designer of this segment of interstate highway laid out the horizontal alignment with a series of straight lines and then connected them with near-maximum degree (minimum radius) curves. In rolling country, the better practice would be first to place the curves so as to "caress" the topography and then add the connecting tangents. Once a horizontal control was set, it appears that the designer attempted to have the vertical profile adhere to the ground line. Many highways, unfortunately, have been designed in such a manner.

Several elements of good design have been defiled (pun intended) in this illustration. On the eastbound lanes a horizontal curve is introduced near the low point of a sag vertical curve; farther in the distance, a horizontal curve is placed near the top of a crest vertical curve in the westbound lanes. Aside from poor aesthetics, daytime travel isn't much bothered by these practices; however, the condition does not hold true at nighttime when headlights' limited range of vision provides the bewildered driver with miniscule notice concerning where the road in front disappeared to.

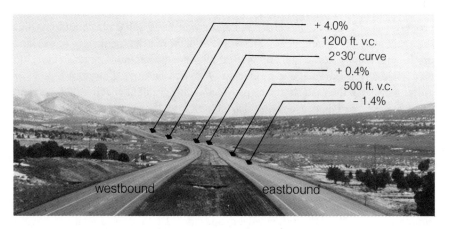

**FIGURE 9-4.** Poor geometry is found on many existing highways.

## What's To Be Done?

When the time arrives to correct this particular section of poor design—maybe after accident records indicate it to be a high-hazard location—the initial fix would be to replace the 2°30′ horizontal curve with a longer radius, perhaps 0°45′, since there appear to be adequate tangent lengths at both ends. A gentler horizontal curve allows more options in placing vertical curves. Going one more step, applying separate control lines for EB and WB lanes through this section would allow for better manipulation of geometry along with more economies in earthwork. There seems no good reason to maintain a single control line on a four-lane divided highway in rolling country.

Consequences of correcting the design:

- Shortened travel distance (approximately 0.05 mile);
- Probable correction of a hazard condition; and
- Long-term cost saving:
  - Travel time and distance,
  - Maintenance effort, and
  - Reduced accidents (maybe).

If corrections had been made prior to original construction, the above savings, together with reduced earthwork, would amply justify the additional thought and effort in working up the preliminary design: another example in the use of Value-Architecture.

## Terrain Considerations

In flat country, highway geometry is rarely dependent on gradient. Land use, property values, waterways, canals, rail and power lines, subgrade conditions, plus the availability of embankment material play the more prominent roles in working up the preliminary horizontal and vertical alignments. Minimum grades enter into play when surface drainage is significant. Water to be carried away in a ditch or gutter parallel to the road should be provided with a gradient of at least 0.5 percent. Financing of road improvements also comes into the picture in mountainous country, where a gradient may be adjusted to balance excavation against embankment to keep costs in line. Caution is advised in cases where the architect may be attempting to economize on earthwork while, at the same time, providing an efficient facility.

Running speeds of passenger cars and trucks on the level are about equal; however, trucks run about 5-percent higher in speed on downgrades and nearly double that in a lowering of speed on upgrades. Undulating gradelines have a consequent effect on traffic operation where there is a high volume of heavy truck traffic. Although an undulating geometry enables trucks to operate at higher speeds when an upgrade is preceded by a downgrade, the condition presents hazards to other traffic. Undulating profiles are suspect, but "roller-coaster" profiles are to be avoided. They generally crop up innocently enough on a long horizontal tangent as a highway profile (designed to minimize earthwork costs) conforms to the natural groundline. Older roads and streets having this attribute are hazardous because hidden dips frequently prevent approaching drivers from seeing each other during passing maneuvers. When the older roads were constructed, the height of the driver's eye was probably high enough to see over the hidden dips; a driver seated in a new low-slung sports car may have trouble seeing over anything higher than an anthill. An existing "roller-coaster" profile can be corrected by prudent use of horizontal curves and/or by employing minor cuts and fills to reduce the grades and increase sight distances.

In general, length of vertical curves ought to fit existing topography, but not be shorter than set minimums reflecting design speed. Over crests, these set minimums have been established from sight distance requirements (as previously stated). Another positive choice, on long grades, is to put the steepest grades at the bottom and lesser grades near the top of the hill. Also, a sustained grade can be broken by short lengths of flatter grade (rather than maintaining a consistent climb) on highways of low-design-speed. However, it should not be overdone: a gradeline having only gradual changes consistent with terrain is better than a gradeline with many breaks between short lengths of grades. In the end, the manner in which the roadway profile relates to the terrain helps to establish the suitability (as well as appearance) of the roadway.

Where intersecting roadways are encountered on steep grades, it is necessary to flatten the profile through the intersection so that both vehicles attempting turns and cross traffic can be accommodated. That is to say, the preliminary profile line should be made as flat as feasible at intersections in order for sight distances and the normal cross section of both roads to remain satisfactory.

## Earthwork

In many large construction projects, the cost of excavating, moving, depositing and compacting soil and rock makes up the greatest share of the total cost. Preliminary geometry is partly an exercise to keep these costs to the optimal minimum.

Rule 1:
    Where no other overriding technical considerations are manifest, keep the road subgrade profile close to the existing ground level.

Rule 2:
    To facilitate a "balanced" job (where excavation and fill are approximately

equal), locate the road profile line so that it continually crosses the existing ground level.

Rule 3:
Ignore Rules 1 and 2 when the Soils Report indicates problems with using existing cut materials for embankment, or if there is a ready market in the neighborhood for clean fill, or if another major excavation nearby requires a waste or spoils area, or if some other unique condition is evident that impacts the economics of the situation.

We'll delve into this some more in the following chapter.

## Additional Considerations

Occasionally unforeseen circumstances arise even when proper geometrics are employed in design. Serious accidents can occur when downhill-traveling trucks and autos encountering moderate curves go out of control as a result of excessive traction forces generated between right- or left-side tires and pavement. Highway furniture (signs, barricades, energy attenuators) as well as temporary conditions (maintenance repairs, local detours) can often thwart the intentions of the best highway geometrics.

Hydraulic mandates for floodplains and river crossings may require adjustments to highway alignment, horizontal as well as vertical. Possible impact to properties resulting from changes in flood patterns (as well as potential risk of flood damage to the highway) needs to be checked out before encroaching upon a floodplain. Depending on hydrologic return period events used for design (10-year flood, 100-year flood, etc.), the highway profile approaching stream crossings will generally be subject to overriding hydraulic considerations. Wherever feasible, floodplain encroachments should be placed to maintain the natural flood flow: direction, quantity, and distribution.

## Preliminary Geometry: Fulfillment

Horizontal control line(s) plotted on the plan view and vertical control line(s) placed on the profile (which also shows the existing ground elevation) are the essentials. Crossroads, interchange configurations, typical section(s), potential extent of excavations and embankments along with a preliminary estimate of major earthwork quantities, approximate right-of-way takings, and a few other items are also desirable at this stage. Enough information should be shown so that, later on, the final geometric design would be little more than "filling in the blanks" and making minor adjustments throughout most of the project length. Preliminary geometry would normally describe tangent bearings, the degree of each horizontal curve, the length of each vertical curve, and some of the more essential dimensions—often only to the nearest foot.

## FINAL GEOMETRY

Once the determinations are made as to materials sources and quality, disposition of water-related elements, pavement design, drainage design and erosion control methods, structure geometry, interchange types, roadside items, and critical right-of-way issues, the final geometry may be computed. Again, we would refer to the AASHTO-90 *Green Book*, but with an added caveat: some of the information contained therein may be suspect.

Sight distance criteria is one concern voiced by a number of *Green Book* critics. Traffic volumes, operating speeds, and vehicle mixes can get extremely complex. AASHTO-90 standards may be based on questionable assumptions regarding vehicle and driver characteristics—very general rather than site-specific. A well-advised practice on the part of the architect would be to check out each situation where sight distance may be in question and determine whether or not braking on a wet surface (plus reaction time) can be accomplished. *Transportation Research Records 1208 and 1303* offer considerable guidance in resolving site-specific sight distances.

Another problem is the unresolved guidance concerning transitions in superelevation between tangents and curves. Should spirals be employed and, if so, how extensive? Should they be used between tangents and very flat curves? How about between compound curves? When designing a railroad alignment, spirals are mandatory, or else a theoretical infinite sideway acceleration is introduced. Vehicles on highways compensate for this by wandering within the lane (the vehicle paths do not travel precisely in the center of the lane), whether or not spirals are used in the geometric layout. Some state highway agencies do not use spirals; others use them to a ludicrous degree. Where not rigidly prescribed, the architect could safely apply the guidance offered in the *Green Book*. However, in the author's experience, the minimum length of runoff—that is, the transition from normal crown to full superelevation—ought to be twice that shown in AASHTO-90, with or without spirals.

Additional concern has been voiced where a highway agency opts for an *e-max* of 0.04 or 0.06 (Tables III-7, III-8, and III-9 in the AASHTO-90 *Green Book*). Recent evidence suggests that the higher centers of gravity of trucks may cause rollovers that result from side friction and the recommended lower superelevations.

Although other complaints against the *Green Book* have been lodged, they do not appear serious enough to override the guidance that AASHTO has developed over the years. The AASHTO-90 *Green Book* remains the basic source for all aspects of highway geometry (allowing for the previously cited caveats), which gets us to the task of finalizing all the geometric elements.

Preliminary geometric patterns (and constraints) have been developed—or may have been established long ago in cases where design is restricted to minor reconstruction of an existing road. Polishing up the mainline profile and horizontal alignment should not, therefore, be an overwhelming chore. Rather, the greater emphasis in establishing final geometry would be directed to modifying intersections, converting intersections to interchanges, removing (renovating) high-hazard sites, and other particulars that would be expected to enhance safety, capacity, and aesthetics.

Where not previously completed, the stationing, bearings, coordinates, elevations, and horizontal and vertical curves, along with ties into existing surveys are, at this stage, computed and checked. (Common practice states that accuracy to 0.01 of a foot (one-eighth inch) is quite satisfactory for most projects.)

## Capacity

Here we get tangled up with "Level of Service" (LOS) and the inordinate complexities of highway capacity. On pages 90 and 91 of the AASHTO-90 *Green Book*, one can find a believable subjective explanation of the six "Levels of Service": A (very good) through F (very bad), followed by (page 92) an optimistic guide for appropriate service levels. Further investigation requires use of the *Highway Capacity Manual*, a guide that requires a great deal of study to apply correctly. Computer adaptations of certain phases of capacity determinations are proliferating, but they must be used with care and prudence.

As most drivers have learned through continued observation, the capacity of almost any urban road or highway is determined by the intersection bottlenecks. Although long stretches of rural roads and interstate highways can provide LOS-A, most design concerns are relegated to eliminating conditions that cause LOS-F ("F" for failure). Logically then, intersection capacities, often determined by operational considerations such as traffic signal phasing and signs, are the bugaboos to be addressed by the highway architect.

Referring to the *Highway Capacity Manual* can be a painful, but necessary, experience if there is likely to be congestion resulting from use of a new or improved highway facility. And the architect is to be reminded that—just as with drainage—consequences both "upstream" and "downstream" of revised traffic flows need to be considered.

## Cross Section

A prevailing tenet in ground transport, be it highway, railway, or pipeline, is consistency of cross section over extended lengths. A pipeline with continually varying diameters—say every hundred feet—would be a most inefficient transport

Geometry    115

vehicle. A rail line in which the width between the rails keeps bowing in and out would be hopeless. Likewise, a constantly varying roadway cross section is not especially functional; neither does a constantly varying cross section lend itself to safe operation. The task of selecting the optimum roadway cross section blends the standard mix of cost, capacity, safety, and efficiency.

## Number and Width of Lanes

To begin, the selection of number of lanes, lane widths, and, when more than two lanes are considered, width and type of median are to be made. Capacity is the primary determiner. Although the *Highway Capacity Manual* is considered the final authority on the subject, many designers have found it (and relevant software) to be overly complex when applied to their circumstances, particularly when dealing with rehabilitation of an existing route under tight financial constraints.

A first cut into capacity says that a two-lane highway can probably handle up to 2,000 vehicles per hour before it becomes a central topic of complaint on the local call-in radio shows. A four-lane divided roadway should be good for well over three times the quantity of traffic of a two-laner. Further refinements are often made by political considerations, which tend to prevail regardless of technological analyses.

Three-lane roads are generally considered a "no-no," except where (in short segments) one of the lanes is for slow-moving vehicles, for two-way left-turn lanes, or for turning storage. If a three-lane facility is selected, the center lane may be a reversible lane, or it may be a continuous passing lane. For the first case (reversible), where commuter traffic travels in one direction in the morning and the opposite in the afternoon, the center lane is usually constructed slightly wider than the outside lanes to allow for portable control furniture (i.e., movable curb, cones, double-line painting). The opposite would be true where the middle lane is set aside for jousting (head-on) and passing. A narrow (but not too narrow) passing lane discourages drivers from continued riding in the center of the road. Passing lanes are addressed in a paper beginning on page 31 of *TRR 1026*. A more complete treatise on the subject by the same authors is found in *FHWA/RD-85/028*. Otherwise, lane widths for any number of lanes are usually in the 3.5 meter range: 11 feet or 12 feet.

Various approaches to the number of lanes and applicable cross sections in suburban settings are presented in *NCHRP Report 282*.

## Shoulders and Curbs

Cursory attention to shoulder geometry can easily lead to a critically weak element in an otherwise well-designed highway. The AASHTO-90 *Green Book*, pages 335-341, contains a brief but good treatise on shoulders. An appropriate clear

116   Highways: An Architectural Approach

(recovery) zone alongside the travel way can go a long way toward enhancing highway safety. Pages 86–95 of *TRR 1122* arm the architect with the rationales behind, and standards for, clear zones. Further discussion on clear zones and the paradox of incipient dangers in shoulders that are too wide can be found on pages 1–7 of *TRR 926* and pages 33–43 of *TRR 806*.

In locations where curb and gutter are to be furnished, the architect must recognize the anomaly that the effective width of the lane adjacent to the curb is reduced by nearly 2 feet in driver perception. The AASHTO-90 *Green Book*, pages 344–351, contains a concise guide to curbs and sidewalks.

## Slopes and Ditches

Drainage requirements once again enter the picture. Soil characteristics and drainage design will, in most instances, dictate the cut slopes, fill slopes, and roadside ditches and channels. Where slope geometry is not specifically dictated by soil and drainage, an opportunity for aesthetic enhancement can be undertaken.

## Standard Cross Sections

Virtually all major highway agencies have standard cross section templates for various classes of highways, quantity of traffic, and topography. The architect is thus relieved of making the selection, but has the responsibility of ascertaining if the predetermined prism is appropriate to the specific facility under design.

## Intersections

Finalizing the geometry of at-grade intersections is one refinement that is too often ignored by many highway designers. Angle (or skew) of intersection, approaching gradients, and fusing of two differing roadway prisms are the primary influences on at-grade intersection geometrics. The AASHTO-90 *Green Book*, beginning on page 669, outlines and explains many of the considerations relative to intersecting roadways.

Occasionally, one encounters a set of plans where the mainline profile matches control elevations at the intersection of the control lines, but ignores the geometric characteristics of the crossroad. The result is either a sometimes horrendous deformation of the intersecting road section or field adjustments that usually require violation of appropriate gradient considerations on the mainline, as displayed in Figure 9-6.

Broad intersections with lots of pavement surface can become drainage nightmares; therefore, some contour grading on the plans of the pavement surface is a

Geometry  117

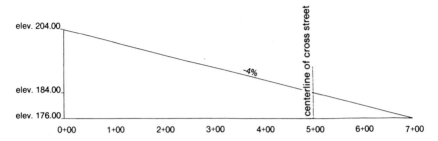

**FIGURE 9-5.** Sometimes preliminary geometry is sketchy. Here is a proposed preliminary profile that meets an existing intersecting street.

**FIGURE 9-6.** If the preliminary profile is not adjusted to match the cross section of the intersecting street during the final design phase, it's up to field personnel to make some adjustments. What usually happens is a minimal modification to the profile in the vicinity of the intersection. Once completed, the intersection might be a great place to film a chase sequence, but the only permanent beneficiary of such poor geometrics would be the nearby service station that specializes in replacement of shock absorbers.

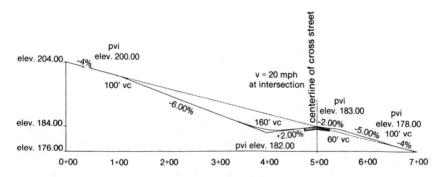

**FIGURE 9-7.** Modifying the preliminary profile on the plans so as to provide a satisfactory transition through the intersection can lead to a safer, more efficient facility. An **appropriate running speed (V)** is the key to establishing lengths of vertical curves. Additionally, the **drainage structures** can be placed at the proper locations.

118    Highways: An Architectural Approach

prudent practice, to prevent the probable proliferation of perplexing problems that panoramic pavement placing present. Gutters and swale channels will not drain very well if the gradient is less than 1.5 percent. A gradient of 1 percent is the minimum for sheet flow on pavements. Not only can drainage patterns be studied, but awkward alignments (i.e., reverse superelevations) can be discovered and corrected when an intersection plan is contour graded.

Another set of ponderables for the architect (and urban traffic engineer) is the tracking patterns of sundry large vehicles, along with their articulated entourages of trailers.

More on intersections, as well as interchanges, can be found in Chapter 13.

### Roadside, Underside, and Overside

The architect must also cope with driveways and other access points. The AASHTO-90 *Green Book*, pages 382–383, is the recommended reference starting point.

Accommodation of utilities within the right-of-way (both transversely and longitudinally) is often desired or mandated. Such a condition often impacts roadway geometry. Mention of and a reference for provisions for utilities is found on page 487 of AASHTO-90 *Green Book*.

Articulating a final design in an urban or suburban setting requires a lot more attention to the smaller details than does a rural design. *FHWA Report No. 80-204* (a fat one) presents a great deal of useful tidbits relating to the urban street design. Although the volume, prepared for an NHI training course, was developed more than a decade ago, most of its contents are still relevant and current. It is not

**FIGURE 9-8.** Where an intersection layout does not provide for the wider turning movements of trucks, the temptation for intimidating actions by truck drivers increases. (King Klang meets Roadzilla.)

indexed, but has a very extensive table of contents. So as to become familiar with the elements presented, an architect might begin with Chapter 17 of the report.

## Pedestrians and Bicyclists

It is understood that foot travelers and cyclists contribute little to highway trust funds and state gasoline taxes (exempting the fact that both groups, at other times, may also be vehicle operators), yet their interests must be considered. The AASHTO-90 *Green Book* outlines, on pages 97–99, 106, and 392–404, a few precautions to be taken during design. More is to be said later in this treatise in Chapter 21.

## Right-of-Way

The last remnant in establishing geometric controls is right-of-way for the highway. Since it could involve access control, surveys, title searches, appraisals, negotiations, relocations, and a number of other facets before all right-of-way issues are resolved, this phase of final geometry is usually left until the completion of all other design activities.

## Final Geometry: Fulfillment

Occasionally, there is concern as to whether the entire geometry ought to be plotted on a sheet separate from much of the topographic and geologic information. Judgment is the guideline to be used: if the plans would be too complicated and difficult to read if all the pertinent information was placed on one sheet, two or more plan sheets of the same area may be used to display different facets of the design. However, for handy reference, either the survey baseline or the geometric controls should be shown on all sheets.

Coordinates, bearings, stationing, horizontal curve data, elevations, grades, vertical curve data, ties, dimensioned cross section(s), and other items necessary to lay out the project in the field should be assembled, tabulated, and mapped. Bridge geometry, positioning of drainage structures, sign locations, property corners, and other items that are not available at this stage are to be included (and tied into roadway geometry) as they become available. It is more than likely that many of the numerical values stamped here as "Final" will still undergo some adjustments before the design is complete.

## Checking

All too often, because of attempts to expedite completion of a project, checking of a designer's work may be performed haphazardly or not at all. Adequate checking of any engineering-related work is not confined to arithmetical computations and/or data processing entries. The rationale, method, approach, and applicability must be reviewed and understood by the persons willing to (or assigned to) stake their professional reputations and personal fortunes behind their signatures as checkers. Although it is the checker who retains the accountability for correctness of the work checked, it is the responsibility of the architect to make sure that the checker is competent and understands the extent of that which is to be checked.

**References**
AASHTO. *A Policy on Geometric Design of Highways and Streets (Green Book)*. 1990. Washington, D.C.: American Association of State Highway and Transportation Officials.
FHWA. *Design of Urban Streets, Technology Sharing Report 80-204*. 1980. Washington, D.C.: Federal Highway Administration.
FHWA. *Passing Lanes and Other Operational Improvements on Two-Lane Highways, Report RD-85-028*. 1985. Washington, D.C.: Federal Highway Administration.
FHWA. *Surface Design and Rehabilitation Guidelines for Low-Volume Roads, Report TS-87-225*. 1987. Washington, D.C.: Federal Highway Administration.
Forgey, Benjamin. 1989. Parkway design a lost art? *Landscape Architecture* 79(3):45–47.
Harwood, Douglas W. 1986. *NCHRP Report 282, Multilane Design Alternatives for Improving Suburban Highways*. Washington, D.C.: Transportation Research Board, National Research Council.
*Highway Capacity Manual, Special Report 209*. 1985. Washington, D.C.: Transportation Research Board, National Research Council.
NCHRP Synthesis 35. *Design and Control of Freeway Off-Ramp Terminals*. 1976. Washington, D.C.: Transportation Research Board, National Research Council.
NCHRP Synthesis 128. *Methods for Identifying Hazardous Highway Elements*. 1986. Washington, D.C.: Transportation Research Board, National Research Council.
Transportation Research Record 806. 1981. *Highway Geometric Design*. Washington, D.C.: Transportation Research Board, National Research Council.
Transportation Research Record 926. 1983. *Traffic Control Devices*. Washington, D.C.: Transportation Research Board, National Research Council.
Transportation Research Record 1026. 1985. *Evaluation Methods and Design and Operational Effects of Geometrics*. Washington, D.C.: Transportation Research Board, National Research Council.
Transportation Research Record 1122. 1987. *Geometric Design and Operational Effects*. Washington, D.C.: Transportation Research Board, National Research Council.
Transportation Research Record 1195. 1988. *Geometric Design and Operational Effects*. Washington, D.C.: Transportation Research Board, National Research Council.
Transportation Research Record 1208. 1989. *Highway Sight Distance Design Issues*. Washington, D.C.: Transportation Research Board, National Research Council.
Transportation Research Record 1239. 1989. *Geometric Design and Operational Effects*. Washington, D.C.: Transportation Research Board, National Research Council.

Transportation Research Record 1244. 1989. *Traffic Grade Crossing Control Devices.* Washington, D.C.: Transportation Research Board, National Research Council.

Transportation Research Record 1303. 1991. *Geometric Design Considerations.* Washington, D.C.: Transportation Research Board, National Research Council.

# Part 3
## Final Design

# 10
# Earthwork

Beginning in the 1950s, when an extensive effort to build new highways across the United States got underway, dirt moving was the largest money item in highway construction work. With more attention now directed towards rebuilding and reconstructing the national highway system, moving humongous masses of soil has diminished in importance. Yet earthwork remains an essential component in roadbuilding, as quality rather than quantity now occupies center court. In its simplest form, earthwork for a road construction project means that soil from excavations (cuts) is placed into embankments (fills) to accommodate the flowing geometry of a highway as it traverses the more uneven terrain of the countryside (or cityside).

## CLEARING THE WORK SITE

On both reconstruction and new construction projects, a certain amount of initiatory work needs to be done in order to build a highway. For urban and suburban locales, businesses and residents may have to be relocated; some utility work must be taken care of; where necessary, houses and buildings are moved or demolished; archeological and other cultural sites are flagged; topsoil is stripped and stockpiled for later use. This accomplished, the next step is "clearing and grubbing," after which temporary ditches and pipes are put in place to accommodate surface drainage during construction activities.

## EXCAVATION

Removing portions of the terrain is considered by most concerned folks as tampering with the natural environment. This is close enough to the truth to require that the terrain removal contribute in a positive manner to the human environment. That is, if the *natural environment* is to suffer, the *human environment* ought to be enhanced to an equal or even greater degree. Excavations for transportation improvements should not be rampant or excessive, for aesthetic reasons of course, but also for important economic considerations (like spending less money). There are a number of different excavation procedures involved, and all types of excavation are not equal.

*Unclassified excavation* is that which forms the bulk of most projects. Traditional practice states that a contract price bid for unclassified excavation includes its transport to and placement in fills, along with appropriate site finishing or trimming.

*Rock excavation* is considerably more difficult than ordinary dirt digging. Most of the time rock excavation necessitates blasting or ripping the rock and then cleaning up the rubble. A technique to obtain a fairly even backslope is known as presplitting, where charges are placed in equally spaced drilled holes at the top of the rock cut. The charges are simultaneously detonated and, if all goes as planned, a fracture line is created from which all remaining rock in the cut can be removed by ripping, additional blasting, or other suitable methods. Many landscape architects (LA) feel that the backslopes should not be presplit evenly on a straight line as the engineers usually direct. LAs suggest uneven presplitting so that the final product will resemble a natural rock facing instead of the rear postern to an unfinished federal penitentiary.

*Trench excavation* for drainage pipes, utility lines, and similar appurtenances requires more specialized equipment and greater care, especially during placement of backfill.

*Structure excavation* has unique concerns in that the soil or rock that remains must have the ability to support considerable weight. Also, quantities excavated must allow for forms and working room around the footings.

*Underwater excavation* is a specialized field that will not be dealt with here. The rules are sometimes strange; cofferdams and caissons may be required; therefore, it is best for the architect to consult with top local expertise: someone with a good track record in developing specifications and estimates for this type of work.

### Embankments

Compaction of embankments plays its part in assuring that a roadway pavement or structure holds up through the expected life of the project. For this reason, embankments of any size are built up in layers and the water content is closely monitored so that it remains close to the optimum (not maximum) value, so as to achieve adequate soil density after compaction (see Fig. 10-1).

Allowing at least one winter to elapse between placing an embankment and constructing the pavement atop the embankment can be very propitious. In the freeze-thaw belt, the repeated working of ice and water usually enhances the soil density. Also, if the construction schedule permits, surcharges placed on top of fills at bridge approaches and left there for a year or two can help provide adequate compaction within the fill after the surcharges are removed. Vertical sand drains are another useful concept in assuring that fills that absorb too much water remain sturdy. Sand drains continually drain away the excess moisture that becomes absorbed by the soggy soil. The Japanese have had success with a variation called sand compaction piles.

In any case, it is prudent when designing a highway embankment to allow for slightly flatter slopes than required so that slides and uncertain angles of repose do not cause unnecessary problems for those who must cope with construction contingencies. If right-of-way is tight, consideration of retaining walls to contain fill within the flatter slopes, rather than attempting to economize with marginally steep slopes, would be the way to go.

Some of the more accepted methods for dealing with problematical embankments can be found in Holtz, 1989.

Borrowing from a source outside of the work area is necessary when there is an insufficient amount of satisfactory excavated material on the jobsite to complete the requisite fills. But, unlike borrowing a neighbor's lawn mower, borrowing good solid dirt is usually expensive. One cost, sometimes overlooked, is the restoration of the borrow pit to a condition equal to, or better than, before the material was taken.

Also expensive is the carting of excavated material to an embankment some distance on the worksite from where it was excavated. When this distance gets to

**FIGURE 10-1.** Embankments to support highways are compacted in layers.

be more than a quarter of a mile or so, a contractor is often paid extra for an item called "overhaul."

Backfill over trenches is a very sensitive issue. Although specifications of most highway agencies are quite detailed in this matter, the control during construction plays the larger role. Transverse trenches underneath pavements cause the biggest migraines. Efforts to address these headaches include using materials other than soil for trench backfill. In Michigan, some success has come through the use of flowable fly ash, a mixture of a small amount of portland cement with type F flyash and water. Sand and flowable mortar are used for backfilling bridges and culverts in Iowa and elsewhere around the country.

## ESTIMATING QUANTITIES

Measurement of excavation is done on a cubic yard basis, except where trench excavations are paid per linear foot. Earthwork quantities are traditionally calculated by averaging sequential (contiguous) cross-sectional areas of material to be excavated (or filled) and multiplying by distances (stations or half stations) between the cross sections. Although an approximation, it is considered "close enough for highway work." Most computer and calculator programs use this approach.

Figuring the major earthwork quantities must allow for swelling and shrinkage. Factoring of the excavated material enters into the calculations so that the quantities of fill material can be adjusted to provide an estimate of some accuracy. For example, 500 cubic yards of excavation may require 600 cubic yards of space in the hauling vehicles and then wind up after compaction as only 480 cubic yards of embankment. Because earthwork is dependent on where the roadway and ancillary structures are to be placed, final geometrics, drainage networks, and other connected features of the project must be designed before final earthwork quantities can be computed.

### Achieving a Balance

To gain the best economic benefit, a balance between cuts and fills is desirable. As discussed in Chapter 9, on geometry, the highway subgrade profile line in rolling or hilly country remains close to the existing ground level, where possible, and it continually crosses and recrosses the existing ground: not remaining above or below. Areas of cut and fill shown on the subgrade profile are about equal. An approximate balance of earthwork is often sought in the early stages of developing roadway geometry. More sophistication is required when the final design is undertaken.

Balancing of cuts and fills is not always appropriate, however. In very flat country, the profile line is usually maintained at an elevation several feet above the existing ground so that the underground water table is not encountered, drainage structures can be accommodated, and snow can be plowed off to the sides of the roadway prism (see Fig. 10-2). Some of the necessary fill material (borrow) needed to keep the roadway profile elevation above ground level may be obtained from drainage channels dug on each side of the road. This explains why, in certain prairie states, the ditches alongside and parallel to the highway are referred to as borrow pits (occasionally corrupted to "barrow" pits). Most borrow pits, though, are found outside of the project limits.

Planning for dirt moving used to rely on hand-cranked preparation of mass diagrams that described the most efficient ways of moving masses of soil from excavations to nearby fills. Nowadays, many highway agencies and their designers rely on computer or even calculator programs to come up with the optimum earthwork strategy, along with corresponding quantities and cost estimates. Therefore, the method used by the sponsoring agency to develop an earthwork balance tends to dominate.

Although a perfectly balanced job would have all the excavated material end up in fill embankments with nothing left over, most of the time some of the excavated material is not satisfactory for embankment. Thus, perfectly balanced jobs are rare. Unsuitable material must be wasted or sold if there is a market available, or it may be used to flatten fill slopes to be incorporated as part of the landscaping scene, or for noise barrier berms. If excess excavation can be sold nearby, the savings may be reflected in a lower bid price. On the other hand, a shortage of suitable excavated material means that a borrow source must be obtained.

**FIGURE 10-2.** Pavement level of a highway in flat open country is placed somewhat higher than the surrounding ground.

## SOIL ON THE WORKSITE

Determining site-specific characteristics and conditions is part of a good design effort. Since earthwork is often a very unpredictable element in any major road improvement, the architect responsible for design must apply the results of sampling and testing outlined in Chapter 5 (under the heading of soil and geotechnics) in order to gain an understanding of properties and behaviors of soils to be encountered.

**References**

Barksdale, Richard D. 1987. *State of the Art for Design and Construction of Sand Compaction Piles*. Washington, D.C.: Department of the Army, U.S. Army Corps of Engineers.

Buss, William E. 1989. Iowa flowable mortar saves bridges and culverts. *Transportation Research Record 1234*:30–34.

Holtz, Robert D. 1989. *NCHRP Synthesis of Highway Practice 147, Treatment of Problem Foundations for Highway Embankments*. Washington, D.C.: Transportation Research Board, National Research Council.

Johnson, Stanley J. 1975, *NCHRP Synthesis of Highway Practice 29, Treatment of Soft Foundations for Highway Embankments*. Washington, D.C.: Transportation Research Board, National Research Council.

Krell, William C. 1989. Flowable fly ash. *Transportation Research Record 1234*:8-12.

TRB State of the Art Report 8. 1990. *Guide to Earthwork Construction*. Washington, D.C.: Transportation Research Board, National Research Council.

# 11

# Drainage and Erosion Control

It has formerly been held as a basic tenet in highway circles that water must always be drained away from highways. This seems to make sense in most instances. Thus, on the average, 25 percent of highway money has been, and is now, allocated for bridges over waterways, culverts, and water-conveying appurtenances: dikes, channels, ditches, velocity controls, storm drains, and sewers.

Nonetheless, there are occasions where the "get-rid-of-all-water" philosophy might well be tempered, if not challenged. It may sometimes be more value-oriented to allow a floodplain the luxury of being inundated periodically. Structures and/or channeling to prevent inundation may be cost prohibitive. Drainage design to prevent overflowing of dikes and embankments is a good idea in arid regions, since destruction of embankments overtopped by floods can be severe in dry climates. Slopes on dikes and embankments in wet climates, on the other hand, tend to be protected, as they are covered by vegetation. Sedimentation basins, as discussed in *NCHRP Synthesis No. 70*, pages 2-34, and more permanent multiple use impoundments may be beneficial to more than just the highway user, even though using a highway embankment as a dam isn't always desirable.

Potential downstream flooding, maintaining water quality, and enhancing visual aspects may require that thought be directed to other than traditional drainage methods. Consequences of limited attention given to drainage during design and construction are often felt later on: Cleaning and repair of improperly designed structures, added to the expense of rebuilding same after unusually heavy storms, can usurp a heavy share of money allocated for maintenance.

To begin, the architect or designer should be armed with the project sponsor's drainage manual (the state highway agency manual would suffice when a local government or private sponsor doesn't have a manual) and a textbook or handbook on the subject for situations in which the drainage design is likely to be complex. (For very simple projects, good judgment is often adequate and more cost-effective than mastering all the intricacies available to the drainage specialist.) In any case, local practice and the drainage procedures of the sponsoring highway agency ought to be followed.

## MAJOR STEPS IN DRAINAGE DESIGN

Following are the major steps in designing the drainage scheme for uncomplicated projects:

- Hydrology—Determine quality and rate of runoff water to be handled.
- Underground water—Ascertain if groundwater conditions are pertinent and significant.
- Hydraulics—Design a system to accommodate runoff and, where necessary, groundwater.
- Legal and environmental aspects—Consider upstream and downstream consequences of the proposed hydraulic design.
- Review—Reexamine the hydraulic design, so as to assure compatibility with the environment, protection afforded the highway, and cost-effectiveness.

### Hydrology

This topic begins with measurement of contiguous drainage areas upstream from the highway location, coupled with the determination as to the return event: how high a runoff one should design for. A "25-year flood" does not necessarily refer to

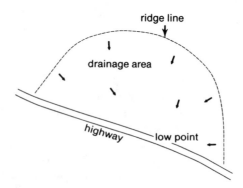

**FIGURE 11-1.** Quantity of water to be carried by a culvert from one side of the highway to the other can be estimated from the size and characteristics of the upstream drainage area.

the maximum runoff that would be expected (based on past experience) in any 25-year period, but rather the expectation that there would be 1 chance in 25 that this high a level runoff would occur in any given year. Hydrology is actually an art form suitable to Las Vegas, Atlantic City, and their oddsmakers. Permutation, combination, and probability are the reckoning parameters; the architect must gamble with Mother Nature.

The result of a hydrological study would be the flow of water that is expected to run off each drainage area upstream from the highway location for the design conditions selected (10-year flood, 50-year flood, etc.) *TR Record 1073*, pages 28-34, compares nine different hydrograph methods and lists a few additional references that may be useful to the designer of the drainage scheme. A reasonably complete treatise on Hydrology and how it relates to highway design is presented in *Hydraulic Engineering Circular No. 19*. Those who desire an understanding of intensity and distribution of rainfall are referred to the four volumes of the FHWA's *Local Design Storm*. Should a designer need only a working knowledge of the subject, the third volume (*User's Manual*, FHWA/RD 82/065) contains both instructional information and a related computer program.

Many highway agencies, as a check and balance, suggest using two or more methods to determine runoff and then selecting one or the other (or an average). However, more and more, the use of computers is relegating the often-subjective art form of hydrology to a scientific, statistically valid branch of applied science. Therefore, if a particular highway design is likely to encounter a significant drainage effort, the investment in appropriate and acceptable software is the prudent route to take.

### Underground Water

Less visible than runoff, but potentially significant, is groundwater. To design, for instance, a depressed section of roadway without first determining water table and ancillary underground aqueous conditions would be foolhardy.

Another situation, also less visible, is the impact of placing a large—or even moderate—highway embankment over an underground aquifer. For years, farmers have complained about the drying up of their lower lands that are near a new major interstate highway complemented by marshes created in the higher land on the other side of the highway embankment. This condition usually results from the

**FIGURE 11-2.** A large highway embankment can interrupt the natural flow of underground water.

weight of the highway fill constricting the natural draining underground water, backing up the flow on the upstream side and lowering the water table on the downstream side. Conventional practice in such cases is to excavate and place drains (transverse to the highway) in the water table before placing the highway fill.

Report No. *FHWA-TS-80-224* is a useful reference should it appear that any phase of underground water is likely to become "sticky." Embankment draining is addressed in *TRR 1159*, pages 39-67.

## Hydraulics

Rural locales dictate an emphasis on culverts, ditches, channels, and dikes. The greater part of the effort by the designer of the drainage scheme is aimed towards getting water from one side of the highway to the other through culverts. Size, type, inlet, and outlet conditions are the design commodities. Ditches (a term that seems out of place when contrasted with other terms of modern technology) are also of certain importance when intercepting perpendicular (and carrying parallel) flows. The most common drainage ditches are discussed and illustrated on pages 26-28 of *FHWA/TS-87/225*. More sophistication can be included, albeit at greater initial cost, by the use of channels and dikes. Less sophistication can be accomplished on low volume roads by providing dips in the profile to accommodate intermittent flows over the top of the pavement surface; initial cost is virtually nil. And, not to forget, attention must be paid to the farmers and their irrigation requirements.

Vegetated ditches can be useful should the water velocity need to be slowed down. Bio-filtration is a desirable spin-off when vegetated ditches are provided downstream from the highway; some of the road pollutants (oils and salts) are intercepted and dissipated before they reach any receiving waters. A maximum gradient of 8 percent is recommended for vegetated ditches. Their length, for bio-filtration to be effective, ought to be at least 200 feet.

Chapter 6 ("Water") dealt with providing adequate fish habitat when relocating portions of streams. The same applies to larger drainage ways that also contain fisheries. There are regulations in many parts of the country requiring that culvert designs accommodate fish passage.

Another refinement to be applied to drainage schemes in rural settings is to lay back the slopes into intercepted brooks, draws, and gullies. Not only does such a procedure help protect against erosion, it also enhances the natural appearance.

A primer for highway drainage in nonurban areas can be found in *FHWA IP-85-15*. In particular, the sections on "Hydrology" (pages 13-18). "Siting of culverts" (pages 18-24), and "Culvert design" with sample problems (pages 25-64) can be very useful for the architect not experienced in hydraulics. Two types of edge drains (for removing water from the pavement base) are described on pages 4-6 of the Oklahoma publication *Pavement Edge Drains*. A brief description of pipe materials is contained in *NCHRP Synthesis 50* (pages 11-14). However, subsequent to publication in 1978, several new coating materials have been introduced and some of the specifications cited have been modified or dropped.

Urban locales compel the expenditure of more money per gallon on drainage facilities for the same purposes. Gutters and storm sewers are expected amenities in a city or, for that matter, suburban environment. Often overlooked in urban designs is the runoff (or lack thereof) on pavements at intersections—particularly those where one or both of the intersecting roads enters at a steep grade and the two cross section templates don't match up too well. A good practice (as mentioned in Chapter 9) is to contour-grade the intersection area on the plans rather than leaving the problem for field personnel. *Hydraulic Engineering Circular No. 12* is a useful drainage design guide for urban and suburban roads.

On city streets, drop inlets or catch basins should be provided upstream at virtually every intersection, to intercept surface water before it has a chance to enter the intersection. The storm drain system for the street must be provided with an outlet—often into an existing storm drain that might already be near capacity. Additional pavement surface reduces the natural ground absorption ability, causing more runoff. What may have been an adequate municipal storm drain system can well be overtaxed when additional drainage requirements are imposed. Consultation and coordination with the local storm water agency must take place prior to dumping any additional water into their storm drains. Widening existing streets may become costly if an extensive storm sewer system has to be built just to carry extra runoff (created by the greater area of pavement surface) to discharge at a considerable distance from the site of the street widening.

Drainage design calculations rarely allow for debris that gets carried down ditches or gutters and becomes lodged at the entrance to a culvert or in a grate over a drop inlet. Clogging of this sort can back up the flow to the extent that pavements

**FIGURE 11-3.** Drop inlets and catch basins are required for urban and suburban street drains.

become inundated and impassable. Although maintenance forces can be expected to keep such debris out of the drainage network between storms, the drainage design must make allowances for the contingency of debris buildup during storms.

When completed and checked, a typically modest drainage scheme may take into account the following:

- Ditches—lines, grades, cross sections.
- Channels—lines, grades, cross sections, lining materials.
- Inlets—catch basins, drop inlets, culvert inlets (all specified as standard types commonly used by the sponsoring highway agency).
- Pipes—sizes, materials, invert (flowline) elevations.
- Junctions—manholes, junction boxes.
- Outlets—end sections, rip rap, leaching basins.

The drainage design is often handled as a separate report containing calculations available to the field supervisor (in case field changes are necessary). A schematic layout shown on the plans may help clarify some questions if there are conflicts with other underground utilities or if below-grade soil conditions (i.e., rock or boggy material) encountered do not allow for installation according to plan.

### Legal and Environmental Aspects

Some of the bloodiest skirmishes in the history of our republic have been initiated over water rights. Nowadays, the battles are raged in courts. Riparian (water) rights research, on a project-by-project basis, is a necessary adjunct to any potential rechanneling or stream diversion that may come about through constructing a new highway or modifying an old one.

Flooding and water quality deterioration are also issues that may have to be addressed and mitigated by the highway architect. Corollary to flooding, legal questions often arise where floodplains are involved. The highway architect must have a "fall-back" position if subpoenas start proliferating in her or his direction. The best position to be in is one of having used a widely accepted methodology in developing the design. For highways in or near floodplains, the FHWA *Hydraulic Engineering Circular No. 17* is a reliable and accepted design guide should there be the possibility that litigation will be brought against the highway agency. The risk analysis procedure is cumbersome, but the FHWA circular presents some example problems, along with requisite documentation, in the appendices.

The U.S. Environmental Protection Agency (EPA), is charged with protecting water quality. Many states have additional requirements that must be followed, especially so if any salt from wintertime maintenance efforts, petroleum residues, and other contaminants have the opportunity to enter the downstream discharges from the highway drainage scheme.

National Cooperative Highway Research Program *Report 264* outlines and describes many of the regulations and pitfalls of constructing highways in the vicinity of defined wetlands.

## Review

Once completed, it is imperative that all aspects of the drainage problems, as well as design of appurtenances to address these problems, are reviewed and checked. Protection of the highway investment both upstream and downstream effects potentially created by the new or modified highway deserve consideration in the review.

Basic factors the reviewer should recognize include:

- Adequacy of the drainage plan in coping with expected conditions;
- Cost-effectiveness of drainage components with relation to the highway investment (i.e., to spend $1,000,000 on drainage appurtenances for a $50,000 highway segment could be considered ludicrous under most circumstances); and
- Suitable harmony with the natural and human environment.

## EROSION CONTROL

Another episode in the suspense-filled saga of highway drainage deals with the prevention of washouts:

- During construction;
- After the highway is opened to traffic; and
- In material sites (borrow pits).

Before tackling any major potential erosion control, the architect is advised to peruse *NCHRP Synthesis 18*. Although a bit dated, it still contains a good deal of relevant guidance. Pages 5-12 describe the initial considerations involved; pages 15-25 outline a few revegetation techniques and related field practices; a glossary can be found on pages 32-38; checklists on pages 39 and 41 can prove to be quite helpful; and Appendix D (pages 42-46) sums up the advantages and problems with various standard practices prior to widespread use of geo-textiles.

### Slopes

Almost all erosion conditions are related to slopes exposed or created by construction activity. Therefore, primary attention must be given to the design of and the treatment applied to these slopes.

Moisture in the soil, moisture moving on the ground surface, and moisture moving through the subsoil are vital factors in designing cut slopes and, to a lesser extent, in fill slopes. A small amount of moisture in sand improves the stability, and moisture is necessary for establishing vegetation and soil stability. Excess soil moisture can cause erosion, surface slippage, and massive slides, which can greatly influence any slope design. If the moisture is concentrated in a definite location or channeled, it can be easily drained away. If the water is distributed extensively throughout the slope and there is enough to create serious problems, some type of drainage system must be used.

Formations from which cut slopes are made and material obtained for fill slopes vary greatly in size, shape, texture, color, moisture content, and material composition. Effective slope design would consider all of these. The usual typical section in highway design has a separate guide for rock, while other materials have a common slope guide. Lately, it has been recognized that all soils may not fit a common repose. A more fitting categorization might be *loose, compressed*, and *cemented*, to describe the natural conditions of soil.

Loose material is composed of soil particles of various sizes that have been bonded together by compaction or cementing. Sand, gravel, and diffused rock are the most common loose materials because they can be bonded only by cementing, a slow natural process. (Clays and silts, also found in a loose condition, are easily bonded by compaction when moist.) Loose material will shift and move when pressure is applied and will not usually repose on a slope steeper than one-on-two (2:1); it may not be stable on flatter slopes. The finer particles are very subject to both wind and water erosion and need a surface erosion control when the natural control is removed.

Compressed material is composed primarily of clay and silt that have been lightly bonded together by compression when moist. Larger sand and rock particles can also be included if there are enough fine particles to act as a bonding agent. Compressed materials will repose nearly vertical but become subject to very rapid breakdown, and, because any material that breaks down becomes loose, cut slopes steeper than one-on-one-and-one-half are not recommended. Paradoxically, a near vertical cut may be more stable than a sloped cut, possibly because the vertical slope does not receive the water penetration and consequent destabilization. Compressed non-vertical cut slopes should, in some climates, have the surface scarified or loosened to provide a planting bed for vegetation, so as to help prevent or allay washouts.

Materials that are stable when cut vertically and not subject to rapid deterioration can be considered cemented. Rock belongs in this group. Rock usually is very stable, but the hardness can vary down to the point where material is only compressed. Because rock is stable and does not deteriorate, about the only way it breaks down is by jarring or by water freezing in the cleavage cracks. Vertical cuts in solid rock make it possible to reduce the amount of excavation needed, requiring less right-of-way and virtually no erosion controls. Yet, even with these reductions in excavation quantities, rock cuts are often avoided because the cost of excavation is high. As rock cuts and those involving other cemented materials will not be covered by topsoil and vegetation, it is desirable to make them function and appear natural by blending in the rock shapes and textures—a difficult task because usually the weathered surface of rock is different in texture. Also, rock that is completely buried cannot be analyzed until the soil around it has been excavated. By studying nearly exposed formations and by drilling, information can be obtained to develop a concept on how a rock cut should be made, but the final surface design will have to be made during construction of the slope because the natural cleavage lines along which most of the cuts should be made are not always known until the rock is exposed and possibly broken into.

## Contour Grading

Since nature does not usually shape the land to meet current geometric roadway standards, some modification of the landscape is necessary to construct a modern highway with its requirements for size, vehicle quantities, and speed. Usually, highway construction is divided into three phases: earthwork, structures (including drainage), and surfacing. Earthwork may be considered the portion of the highway construction necessary to provide a strip of terrain in the landscape on which the structures and road (surfacing) can be placed. This reshaped strip of land will have the width, grades, compaction, drainage, and so on, to accommodate the highway, but, outside the highway limits, grading must be related to the adjacent terrain and will not generally be controlled by highway geometrics.

Natural ground surface tends to be irregular, random, and continually changing. A highway, conversely, has smooth parallel edges and remains the same for long distances. Reshaping of the land ought to change from the road edge character to the undisturbed natural. Contour grading is a means to locate, design, and construct cut and fill slopes that will aesthetically accomplish such a transition. The architect may employ the tool of contour grading as the first step in controlling erosion, strongly suggesting that initial prevention is better than applying corrective measures later on. Additionally, enhancement of many aesthetic characteristics can be achieved through well thought out grading plans.

## Rills, Gullies, and Sheets

Even when the best slope design is put forth, there will be some continual soil loss on cuts and fills. Development of rills (small eroded channels) and gullies (slightly larger erosion channels) do occur. The amount of sheet and rill erosions can be guesstimated, if the problem has incipient questions, by applying the U.S. Department of Agriculture Universal Soil Loss Equation (USLE) in order to determine if there is an appropriate vegetive cover to protect the cuts and fills. The procedure is complex and is only justified for projects where large areas of erosion are foreseen.

Crown ditches at the tops of slopes are often required to intercept flows coming down the hillsides before the the flows have a chance to form gullies and erode slopes. The crown or interceptor ditches convey the water either to the ends of the cut slopes or to channels (usually paved or rip rapped) leading down the slopes. Additional right-of-way acquisition may be required, to allow for crown ditches and other economical erosion control measures.

### Recycling

We keep coming back to the topic of old tires. Discarded rubber tires can be used effectively for temporary shoulder stabilization (up to ten years) and for slope protection. The key is to have the tires properly installed and adequately anchored. Although not especially aesthetic, they are cheaper than the more conventional materials and are certainly viable in locations where their visibility is restricted.

### Construction Activities

A great deal of erosion occurs during construction activities, especially where work by many subcontractors is not well coordinated. The architect is urged to establish strong controls (via specifications and special provisions) over contractor activities so that field personnel and inspectors have suitable leverage to prevent, alleviate, and/or lessen this phase of potential erosion.

### Vegetative Cover

Vegetation can be an important protagonist for cut and fill slopes. It provides additional texture and variety for most of the visual characteristics of the landscape, but is even more important for erosion control and slope stability. Vegetation takes time to replace after being removed; cutting into a hillside that is covered with large trees will create a scar that will remain visible for a long time. A grass community may be replaced in a couple of years. Grass is about the quickest natural means to provide erosion control, but might not be compatible with other vegetation. Trees can be very effective for screening, delineating, accenting, and bordering, but they are often restricted because of narrow right-of-way and safety clearance. Shrubs are very effective for roadside planting and have versatility around other objects. Nevertheless, availability of adequate water—either via normally high rainfall or irrigation—is mandatory for successful plantings.

### Geo-textiles, Fabrics, and Membranes

A somewhat redundant or overlapping (unintended pun) title, this suggests that there are a number of materials useful for erosion control, each for a specific condition. Since the development of new materials and their application is changing at an adventurous pace, any specific suggestions offered here would be obsolescent by press time (and fully obsolete by the time the reader gets to this chapter). The architect is advised to check with the most up-to-date directives dealing with geo-textiles. Also helpful are material vendors, with the understanding that they may not be entirely objective.

Geo-textiles come both woven and nonwoven; most are porous, some are not. Porous fabrics are used on slopes and as components of retaining systems. Membrane liners generally are impregnated with a bituminous compound to prevent seepage of water. They are most often associated with ditches, channels, and pondings.

Properties of geo-textiles must be tied to the function they are to perform. Installation during construction ordinarily dictates strength requirements. What with heavy equipment rolling over and the stretching and pulling necessary to install and overlap properly, fabrics may be put to their greatest stresses before they are required to perform their function.

## DRAINAGE AND EROSION CONTROL: FULFILLMENT

Design of the drainage scheme can be considered to be complete when the following have been accomplished:

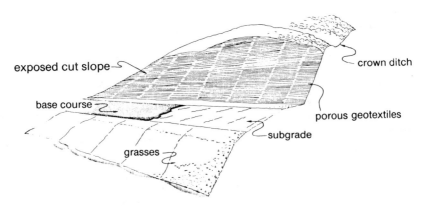

**FIGURE 11-4.** Options are available to control erosion from exposed cut slopes.

142    Highways: An Architectural Approach

- All hydraulic calculations and results in an understandable format. It may be necessary for some future adjustments in the field, and the original design might have to be altered (by someone else).
- Mapping of the entire drainage scheme. Stations, invert elevations, pipe schedules, connections to existing systems, box culverts, stream rechanneling, ditches, irrigation, and sundry other items necessary for the satisfactory installation and operation of the complete drainage scheme.
- Specifications for nonstandard items.
- Estimates (usually tabulated) of lengths, quantities, and so on, of all components of the drainage requirements for the proposed highway segment.
- Upstream and downstream consequences that may be anticipated.
- Schematics where appropriate.
- All aspects of the erosion control measures, with adequate descriptions for field installation. This would include estimates of quantities, specifications for specialty items, locations, mapping, stationing, and cross sections.
- CHECKING EVERYTHING STATED OR IMPLIED IN THE ABOVE!

**References**

AASHTO. *A Guide for Transportation Landscape and Environmental Design.* 1991. Washington, D.C.: American Association of State Highway and Transportation Officials.

Berthelsen, Gene. 1989. Erosion control. *AASHTO Quarterly* 68(2):6–7.

FHPM 6-7-3-2, *Location and Hydraulic Design of Encroachment on Floodplains.* Regularly updated. Washington, D.C.: Federal Highway Administration.

FHWA. *Geotextile Engineering Manual*, Report TS-86-203. 1986. Washington, D.C.: Federal Highway Administration.

FHWA. *Geotextile Specifications for Highway Applications*, Report TS-89-026. 1989. McLean, Virginia: Federal Highway Administration.

FHWA. *Highway Subdrainage Design*, Report TS-80-224. 1980. Washington, D.C.: Federal Highway Administration.

FHWA. *Hydraulic Design of Highway Culverts*, Report IP-85-15. 1985. Washington, D.C.: Federal Highway Administration.

FHWA. *Local Design Storm*, Report RD-063, 064, 065, 066. 1982. Washington, D.C.: Federal Highway Administration.

FHWA. *Surface Design and Rehabilitation Guidelines for Low-Volume Roads*, Report TS-87-225. 1987. Washington, D.C.: Federal Highway Administration.

Gray, Donald H., and Andrew T. Leiser. 1982. *Biotechnical Slope Protection and Erosion Control.* New York: Van Nostrand Reinhold.

Hydraulic Engineering Circular No. 12. 1984. *Drainage of Highway Pavements.* Washington, D.C.: Federal Highway Administration.

Hydraulic Engineering Circular No. 17. 1981. *Design of Encroachments of Flood Plains Using Risk Analysis.* Washington, D. C.: Federal Highway Administration.

Hydraulic Engineering Circular No. 19, 1984. *Hydrology.* Washington, D.C.: Federal Highway Administration.

Innovation, Research Bulletin 6. 1990. *Improved Construction Erosion and Sediment Control*. Olympia: Washington State Department of Transportation.

Oglesby, Clarkson H., and R. Gary Hicks. 1982. *Highway Engineering*. New York: John Wiley & Sons.

NCHRP Report 264. 1983. *Guidelines for the Management of Highway Runoff on Wetlands*. Washington, D.C.: Transportation Research Board, National Research Council.

NCHRP Report 290. 1987. *Reinforcement of Earth Slopes and Embankments*. Washington, D.C.: Transportation Research Board, National Research Council.

NCHRP Synthesis of Highway Practice 18. 1973. *Erosion Control on Highway Construction*. Washington, D.C.: Transportation Research Board, National Research Council.

NCHRP Synthesis of Highway Practice 50. 1978. *Durability of Drainage Pipe*. Washington, D.C.: Transportation Research Board, National Research Council.

NCHRP Synthesis of Highway Practice 70. 1980. *Design of Sedimentation Basins*. Washington, D.C.: Transportation Research Board, National Research Council.

*Pavement Edge Drains, A Comparison*. 1987. Oklahoma Department of Transportation.

Transpiration Research Record 1073. 1986. *Hydraulics and Hydrology*. Washington, D.C.: Transportation Research Board, National Research Council.

Transportation Research Record 1159. 1988. *Prefabricated Vertical Drains and Pavement Drainage Systems*. Washington, D.C.: Transportation Research Board, National Research Council.

Transportation Research Record 1231. 1989. *Analysis, Design, and Behavior of Underground Culverts*. Washington, D.C.: Transportation Research Board, National Research Council.

Transportation Research Record 1248. 1989. *Geosynthetics, Geomembranes, and Silt Curtains in Transportation Facilities*. Washington, D.C.: Transportation Research Board, National Research Council.

# 12

## Structures

What ever happened to the bridge architects? Wasn't bridge architecture once a recognized, respectable profession? Could not John Roebling, Othmar Ammann, and David Steinman have been considered bridge architects? Highway bridges are now designed almost exclusively by structural engineers. Is it because structural engineers can command higher salaries than bridge architects, so that the latter now prefer to be known by the former designation?

Bridges and retaining walls produce an emphatic visual effect. Not only can they be imposing, but, hopefully, they should be durable: 50 years or so. An architect must help to assure that they will be aesthetically compatible to both the highway and its surroundings. We're not talking here of ornamentation but rather, as stated in the *AASHTO Guide for Transportation Landscape and Environmental Design*, "attention [be] given to scale, proportion, form, line, texture, color...as well as pure function." These factors apply in the initial concept and in the consideration of various alternative types of structures. A highway architect must grab the initiative and take an active, partisan interest in the concept and overall design of the structures that are to be incorporated into a highway segment—not always easy because of the traditional design setup in most highway agencies.

Bridges and other edifices of the highway are almost always handled by design crews who specialize in the field. Responsibility of the highway designer has always been to provide the structural designer with such items as roadway geometries, hydraulic information (where waterways are involved), railroad requirements (where germane), and (oftentimes) soil information for foundations.

## MAJOR TYPES OF MAJOR STRUCTURES

Structural engineers classify bridges according to their means of support and materials of construction. The highway architect would probably prefer a classification more suited to roadway requirements. With this preference in mind, let's suggest a sophisticated high-tech bridge classification system of use to the highway architect:

1. Small
2. Medium
3. Large

### Small Bridges—Box Culverts, Pipe Arches, Slabs, and Rigid Frames

The recently completed *National Bridge Inventory* (*NBI*) catalogues bridges having spans of more than 20 feet. Sometimes, though, even shorter spans are found under highway structures. Small bridges rarely exceed a span of 80 feet.

*Box culverts* made of concrete poured in place as a single unit and capable of supporting heavy loads are constructed as short spans around drainage channels and farm road crossings. Typical length would be 40 feet with a span of 12 feet. They may be multi-barreled where drainage conditions generate large, variable, and/or seasonal flows.

*Pipe arches* are Paul Bunyan's version of metal pipe culverts (see Fig. 12-1). They can often be used in the same situations as box culverts.

*Slabs* (deck units), either precast or poured in place, are designed so that they are capable of carrying large volumes of traffic (and even heavy loads) to span short

**FIGURE 12-1.** Metal pipe arches can be incorporated into attractive bridges over smaller waterways. (Photo: Contech Construction Products, Inc.)

distances, usually less than 30 feet. Support for the decks is provided by abutments and, if more than one span is required, by standard bents or piers.

*Rigid frames* are different from slabs in that the deck and supports are poured in place as a single piece. Rigid frames were built extensively during the 1940s and 1950s. More recently, prefabricated and precast structures have been placed on the market. Comprised of corrugated steel or corrugated aluminum or reinforced concrete, they range in spans up to 40 feet and have the advantage of quick installation so that maintenance of traffic becomes much less of a problem.

## Medium Bridges—Girders and Trusses

Bridge designers prefer to employ the label "stringer" for what a layman terms "girder." Since we are not pretending to be structural designers, we'll stick to the more common term.

*Girders*, comprising more than half of the 600,000 highway bridges in the United States, come in a number of styles using one or more of three different materials for the span: timber, concrete, and steel.

Old timber girder bridges account for some 10 percent of existing highway spans. They are now used more sparingly for unique situations.

Concrete can be incorporated into girders by two methods: box and pre-stressed. Box girders capable of carrying large volumes of traffic are most appropriate where curved roadway alignment is used. Gaining in popularity are segmental, posttensioned box girders. Concrete modules are precast and erected and the reinforcing steel cables are placed in tension. Erection is often done by cantilevering the sections from each bent (see Fig. 18-1). The balanced cantilever method of erection and joining successive segments became popular in Europe during the 1950s. It has been applied sparingly to highways on this side of the Atlantic, the Linn Cove Viaduct in North Carolina and Glenwood Canyon (I-70) in Colorado being notable exceptions. One important advantage of the balanced cantilever is the obviation of providing ground level access and form support along an entire route. The only ground disturbance (or waterway encroachment) that needs to be made is at each pier; although 400 feet is typical, piers can be placed as far apart as 750 feet. Minimal disturbance to the terrain becomes an environmental plus. The big problem with this construction method is the continuing structural analyses that must accompany each step of the erection. Pre-stressed and posttensioned concrete girders take advantage of the principle that concrete is at peak performance while in compression and steel is at peak while in tension.

Steel girders come into play for spans longer than optimum for concrete (see Fig. 12-2). Steel members can be connected to form a continuous beam; longer span structures of steel beams tend to be more economical and more easily constructed than those of concrete. Viaducts are usually comprised of contiguous girder spans.

**FIGURE 12-2.** One of the most common highway structures is the steel girder.

Orthotropic and exodermic concepts incorporate the deck and girders into a single unit so that the deck adds its own component to the support. Although expensive to construct, these designs are useful in strengthening existing bridges and in new structures where dead load is a major concern.

*Trusses* were used extensively for railroad bridges and many highway bridges a century ago. They are constructed as a framework of members set in triangles. Since each member is dependent upon the compression or tension of all the others, one weakened member can result in the complete collapse of the structure. Truss bridges have been replaced by other types of structures for safety reasons, primarily because the members cause obstructions above the road level: a vehicle striking one of the members may render the structure unsafe. Where truss construction is used on a newer bridge (e.g., approach span to a suspension bridge), the truss is sometimes placed below the roadway level.

## Large Bridges—Arches, Suspensions, Cantilevers, and Cable-stays

Once it becomes necessary to put in a major bridge crossing—usually under, rather than over, the highway—emphasis shifts to the requirements of the bridge rather than of the highway. The bridge designer then occupies the catbird seat and may (not always intentionally) present the highway architect a new series of changing and interacting problems that throw the design schedule into the dumpster. One overriding cause for consternation has been that since 1978 Federal Highway Administration requirements, detailed in *Technical Advisory T5140.12*, mandate that alternative steel and concrete designs be considered on major, federally funded highway bridges.

*Arches* can come in all sizes. They may be built of masonry (not likely for new bridges), concrete, or steel. Poured in place, concrete arches are still viable for certain situations. Since concrete is at peak performance while in compression, the structure stands by itself, requiring minimum reinforcement. Capable of carrying large volumes and heavy loads, concrete arches are constructed over washes or rivers; however, they are not generally practical over roadways because of tight clearance through the arch. Typical span is not likely to be much over 100 feet. (Perhaps a concrete arch should not be considered a "large" structure.)

A steel arch may be constructed as a hinged arch that is later (at the completion of the arch) fixed by rigidly anchoring the joints to abutments and adjacent members. Capable of carrying large volumes and heavy loads, steel arch bridges are used where a long span for crossing canyons, gorges, or rivers is required (see Fig. 12-3). Typical spans may reach more than 500 feet. Three varieties of steel arch construction include:

1. Spandrel or standard arch—Roadway is supported above the arch by spandrel columns.
2. Suspension arch—Roadway deck is suspended from the arch by steel cables.
3. Through arch—Roadway deck is positioned halfway between the top and the bottom of the arch; a combination of suspension cables and spandrel columns support the roadway deck.

The latter two have the same caveat that applies to trusses: an errant vehicle can damage one or more the suspending cables and render the entire bridge unsafe.

*Suspensions* hold all records for longest spans. They are generally an entire highway segment in themselves; thus, the highway architect may be in the position of only providing geometry for the approaches.

*Steel-framed cantilevers* are becoming obsolete, but, as already described,

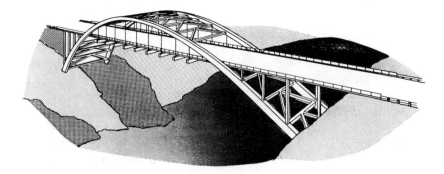

**FIGURE 12-3.** Steel arch bridges are one favored way to cross steep gorges.

cantilever-type construction methods are still in favor where access to the ground or waterway below the structure is difficult.

For spans greater than 700 feet, *cable-stayed* bridges are getting to be popular; they are graceful, even gossamer. Interestingly, cable stays on bridges date from 1817. In both Europe and Latin America, cable-stayed bridges have become a viable surrogate for the more prosaic suspension types.

## INNOVATIVE APPROACHES

Europe has been far ahead of the United States in developing new means, methods, and materials for incorporation into new major structures. The philosophy behind European innovation is partly a result of combining the design and construction responsibilities into one package. The *design and build* procedure looks into types, cost, time, and methods of construction. The designer winds up with both responsibility and accountability. In the United States, the adversarial philosophy dominates. The highway agency, along with the designer (who is often a consultant to the agency) and the contractor, form a poorly harnessed troika that tends to pull in three different directions. Legal and insurance professionals soon find themselves in the contest to the extent that any attempts at innovation may well wind up in litigation.

## ADDITIONAL STRUCTURAL ELEMENTS

### Temporary Bridges

There are times when temporary spans or crossings must be provided while a replacement bridge is under construction. Worthy of mention for this purpose are some items that have been developed by the military:

- Modular systems;
- Pontoon bridges (for crossing docile waterways); and
- Prefabricated.

All these types can be dismantled, moved, or disassembled after they have served their purpose and be used again at some other location, a consideration that allows for somewhat lower costs for each temporary installation.

### Specialized Functions

Highway architects may find themselves faced with crossing situations that require special attention. For instance:

- Overhead flumes for canals, pipelines, or slurry transport;
- Under-highway siphons; and

- Pedestrian overpasses (underpasses are now discouraged) (see Fig. 12-4).

Generally, specialties such as these are handled as modified versions of standard highway structures.

### Tunnels

Services of specialists in this field are usually required to aid the highway designer at times when tunneling becomes the most feasible (or the fall-back) alternative. Should this course of action be the only option, one must be cognizant of several undersirable operational and maintenance features introduced by a highway tunnel: forced ventilation, continuous dewatering, potential fire danger, lighting, and the inability of motorists to adjust rapidly to the difference in light intensities during daytime.

## REPLACING VS. RESTORING

As a result of the *NBI* it has been determined that half of all the highway bridges in the United States are in need of rehabilitation or replacement. Before deciding on removal and replacement of an existing bridge, there should be some determination as to whether an action of such nature is warranted. Can an existing bridge be salvaged by strengthening and/or widening? Or can it be maintained for travel in one direction? Cost is the determining factor, but the highway architect is not out of line in urging the structural designer(s) to examine the possibilities of salvaging an existing highway bridge for continued service.

Several techniques have proved successful in strengthening existing bridges. Adding or replacing structural members can increase deck capacity and reduce the

**FIGURE 12-4.** A through arch bridge may even be used as a pedestrian overpass.

magnitude of the loads placed on existing components. Replacing a concrete deck with a lighter weight deck can reduce the dead load on a structure. Going one step further, providing composite action between deck and girders, as in orthotropic design, can be cost-effective.

Other possibilities include building up existing members with additional plates, posttensioning concrete members, providing additional supports to reduce span lengths, strengthening critical connections, and increasing transverse stiffness. The latter two are considered secondary measures that complement other procedures. Whatever fixing up is decided upon should be undertaken only after a thorough bridge inspection (described in Chapter 22) is completed.

Our society has long been conditioned to the philosophy of rapid (many times instant) obsolescence so that any considerations related to salvage are given miniscule consideration. Structural engineers, human as they are for the most part, would rather design new bridges than fix up old ones. Therefore, a little pressure from the highway architect might be necessary to assure that a needless waste of taxpayer dollars is not incorporated into an otherwise economical highway design. One aid that the highway architect can enlist is the *NCHRP Report 293* (along with the references contained on page 3 therein). The highway architect might wave the report in the face of the structural designer in order to get a reaction.

In spite of what went on in the preceding paragraphs, salvaging or strengthening an existing bridge is appropriate perhaps less than 20 percent of the time (a .200 batting average). The point to be made is that restoring an existing bridge should be given adequate consideration prior to any decision to build a new one in the same place.

## PROBLEM AREAS

Approach lanes to overpass structures have been and continue to be bugbears. Sometimes, dips in the pavement at each end of a bridge deck become evident one or two years after a bridge is opened to traffic. Although this phenomenon does not occur on all approaches, it does happen "too often for comfort." The cause appears to be four-fold:

1. Time required to consolidate the approach embankment and its foundation;
2. Restricted area for compaction equipment to operate next to abutments;
3. Soil erosion at the abutment face; and
4. Inadequate provisions to drain the embankment behind the abutment.

The issue of overpass approaches is addressed in *NCHRP Synthesis 159*.

Scour vulnerability has recently become a hot topic as a result of several major bridge-over-waterway failures. Hydraulic erosion of the soil supporting the footings tends to reduce their load carrying ability, eventually to the extent that when a heavily loaded truck comes along—thump, splash! The FHWA *Hydraulic Engineering Circular 18* presents procedures for designing new, replacement, and rehabilitated bridges to resist scour.

Another item to be considered pertains to bridge railings. Many existing bridges are having or have had their rails retrofitted to conform to FHWA standards. This operation, which is quite expensive, has taken finances that otherwise would have been put toward rebuilding, replacing, or strengthening marginally functioning structures.

## RETAINING WALLS AND SYSTEMS

Right-of-way requirements, stream encroachments, embankment instability, and several other conditions often decree that retaining walls or surrogate measures be employed to contain and/or protect the highway. There are several concepts in use to accomplish these ends:

- Cantilever walls;
- Mechanically stabilized embankments;
- Gravity systems;
- Reinforced slopes; and
- Root pile systems.

A *cantilever wall* resists earth pressures via the bending moments within the wall. Placement of the footing to resist overturning usually requires a considerable amount of extra excavation. This has been the traditional method for protecting roadways in tight cuts and for containing high fills, but rigid cantilever walls can become prohibitively expensive.

*Mechanically stabilized* and *reinforced embankment* techniques are less costly. Reinforcing strips attached to the facing are laid in the soil so that the backfill holds them in place. Geo-textiles and certain recyclable materials, such as discarded automobile tires, are often employed in these types of retaining structures. Tie backs are protected steel cables connected to ground anchors that are anchored in stable soil behind the backfill. The tie backs are used to help support a continuous—as opposed to composite or segmented—wall.

*Gravity systems* come in assorted sizes and shapes. They use the weight of the wall itself, sometimes accompanied by beveling or leaning the wall back. Bin-type and crib-type walls are built up, with members forming bins or cribs that are then backfilled to provide the mass necessary to resist the overturning moment of the embankment (see Figs. 12-5, 12-6, and 12-7). By stepping or terracing, they create opportunities to incorporate plantings and other landscape amenities.

*Reinforced slopes* are coming into vogue with the adaptation of various geo-synthetics. While not actually retaining earthwork, they do permit use of

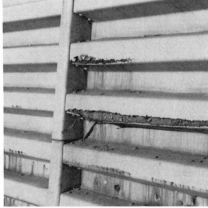

**FIGURE 12-5 & 12-6.** Metal bin retaining walls are suitable in many instances...

...but tend to disintegrate.

steeper slopes that may accomplish the goals stated above. Geo-textiles are not the only materials used to reinforce slopes, however. Gabion walls, mentioned in Chapter 6, are made up of rock-filled metal cages. Placed on the face of a fairly stable embankment, they offer little structural value: their strong point is in resisting weathering or waterway action on the face of the embankment. Rockery can come in many forms and requires some of the same caveats required of gabions. For smaller gravity-type walls, a viable recycling effort is the use of old broken up curbs and sidewalks. Selectively placed so that the broken edges form the facing, they can produce the same effect as ashlar masonry, with proper care in positioning and placing. Also, they can be quite inexpensive if the material is available nearby.

**FIGURE 12-7.** A concrete crib wall is more durable.

*Root piles* are neither retaining nor containing, but an adaptation of a technique used by architects for many years. Occasionally used to widen existing roads on fills, root piling consists of a series of small diameter holes drilled 30–60 feet deep and filled with grout around reinforcing rods (see Fig. 12-8). Interlaced by angling, they are tied together at the top with a concrete cap. The concrete cap forms the sub-base for the widened roadway.

Retaining walls have other uses as well as those described. They have been employed as rock catchers at the base of unstable slopes, to prevent rocks from rolling onto the roadway after they peel off the hillside. Retaining walls are also utilized as slide suppressors embedded in failed slopes, so as to remedy the slope failures.

Some of the retaining techniques mentioned require the service of specialists; others can be adequately formulated by an experienced highway design crew. Aside from guiding the aesthetics, the highway architect can do yeoman service by assuring that adequate drainage systems are employed to prevent moisture buildup in the retained backfill.

## INITIAL DATA REQUIRED BEFORE A STRUCTURE IS DESIGNED

Responsibilities of the highway architect and design crew to provide information to the structural designers are many and varied. Much of the preliminary and some of the final highway design must be completed before the bridge engineers can get moving on their tasks. On the next two pages are some of the items that may need to be provided.

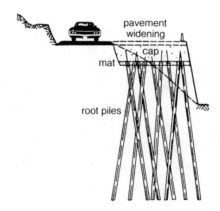

**FIGURE 12-8.** Root piles are an adaptation of a system used by traditional architects for many years.

## Typical Cross Sections

Geometrics developed previously (Chapter 9) would determine the following:

- Cross section of roadway carried by structure;
- Cross section of roadway, stream, railroad, or other facility being crossed;
- Location of control and profile lines, as they relate to cross sections;
- Design speeds for each alignment; and
- Clear zone under structure.

## Plan Sheet

Completion of final geometrics precedes establishment of the following:

- Station ties to control lines at points of intersection.
- Horizontal alignment for roadway carried by the structure (if applicable): bearings, curve data, coordinates of PI's, and stationing.
- Horizontal alignment for roadway or facility to be crossed: bearings, curve data, coordinates of PI's, and stationing. (Ties to existing roadways, railroads, or streams and any physical characteristics that may have an effect on the geometry of the structure and the approach fills may require field surveys to obtain adequate information.)
- Skew (crossing angle) of facilities crossed, unless one of the alignments is on a curve.

## Profile Sheet

These elements furnish the necessary third dimension:

- Profiles of facility carried by structure;
- Profiles of facility being crossed; and
- Vertical curve lengths (checked to make sure they are adequate for stopping sight distance requirements).

## Stream Crossings

A major structure may require that the highway architect (or squad hydraulics engineer) supply most, or all, of the following information on a hydraulics data sheet:

- Drainage area;
- Design flood ($Q_d$);
- 100-yr. flood ($Q_{100}$);
- Normal depth (dn) for $Q_d$;

- Normal water surface elevation for $Q_d$;
- Backwater for $Q_d$;
- Backwater elevation for $Q_d$;
- Velocity through bridge opening for $Q_d$;
- Normal water surface elevation for $Q_{100}$;
- Backwater for $Q_{100}$;
- Backwater Elevation for $Q_{100}$;
- Overtopping flood frequency;
- Magnitude of overtopping flood ($Q_{overtopping}$); and
- Water surface elevation for $Q_{overtopping}$.

**Retaining Walls**

Usually the highway designer provides subsurface as well as geometric details to the structural designers:

- Profiles of the top and bottom of the walls;
- Offsets and stations from the control line to the walls;
- Location, elevation, and size of pipes, sleeves, and other items that go through the walls;
- Cross sections at wall locations; and
- Soils investigations that are applicable.

# GLOSSARY

It may, to some, appear overly ingenuous to define various parts of a bridge in simple terms. Yet the following glossary can be helpful to those highway architects who may be uncertain as to just what an intimidating bridge designer might be referring.

Abutment—The end support for a structure, usually placed in earth or rock, consisting of a footing with earth-retaining backwall and wingwalls.
Bent—Another term for pier.
Clear span—The distance, face to face, between two adjacent supports.
Deck—The roadway surface of a bridge.
Joint—Connection point between two structural members, with or without provision to accommodate differential movement due to thermal expansion and contraction.
Parapet—The low wall placed along the sides of a bridge roadway to prevent vehicles from running off the edge.
Pier—An intermediate support for a structure, usually consisting of a footing, column(s), and a cap to provide support for main structural members.
Span—The distance, center to center, of supports for a beam, slab, truss, or girder.
Spandrel—That portion of an arch bridge above the arch and below the deck.

Stringer—A longitudinal member carrying the deck of a bridge between supports.
Wingwall—The retaining wall extending back from the abutment.

## IN CLOSING...

Needless to point out to the reader, the entire subject of highway structures has been gone over lightly—omissions run rampant. Assuming that design of structural edifices and elements is handled by persons or consultants outside of his or her crew, the highway architect needs to coordinate the design of highway with structures and provide to the structural engineer(s) the following items, in addition to those that were mentioned in the first three paragraphs of this chapter:

- Design schedule;
- Plan sheet format consistent with the highway plans:
- Standard specifications being used on the highway project;
- Estimating forms (blank); and
- Up-to-date structural (or bridge) design manual of the highway agency for the state in which the highway is to be constructed, in cases where the structural engineer(s) are not familiar with local practices.

In turn, the structural designer(s) would be expected to return, in timely fashion:

- Preliminary designs;
- Final designs signed and properly checked by the person or persons qualified and certified to design structures in the appropriate state;
- Special specifications;
- Special contract provisions;
- Estimates of quantities; and
- Anticipated range of bid prices.

**References**
AASHTO. *A Guide for Transportation Landscape and Environmental Design.* 1991. Washington, D.C.: American Association of State Highway and Transportation Officials.
AASHTO. *Manual on Subsurface Investigations.* 1978. Washington, D.C.: American Association of State Highway and Transportation Officials.
FHWA. *Geotextile Engineering Manual, Report TS-86-203.* 1986. Washington, D.C.: Federal Highway Administration.
FHWA. *Hydraulic Engineering Circular 18, Report IP-90-017.* 1991. McLean, Virginia: Federal Highway Administration.
FHWA. *Manual on Design and Construction of Driven Pile Foundations, Report DP-66-1.* 1985. Washington, D.C.: Federal Highway Administration.

FHWA. *Reinforced Soil Structures, Volume I: Design and Construction Guidelines, Report RD-89-043.* 1989. Washington, D.C.: Federal Highway Administration.

Gray, Donald H. and Andrew T. Leiser. 1982. *Biotechnical Slope Protection and Erosion Control.* New York: Van Nostrand Reinhold Company.

Johnson, Stanley J. 1975. *NCHRP Synthesis 29, Treatment of Soft Foundations for Highway Embankments.* Washington, D.C.: Transportation Research Board, National Research Council.

Murillo, Juan A. 1988. Modern bridge construction and engineering services. *Managing Innovation: Cases from the Service Industries.* Washington, D.C.: National Academy of Sciences.

NCHRP Report 290. 1987. *Reinforcement of Earth Slopes and Embankments.* Washington, D.C.: Transportation Research Board, National Research Council.

NCHRP Report 293. 1987. *Methods of Strengthening Existing Highway Bridges.* Washington, D.C.: Transportation Research Board, National Research Council.

Saito, Mitsuru, and Kumares C. Sinha. 1990. Timing for bridge replacement, rehabilitation, and maintenance. *Transportation Research Record 1268*:75–83.

Technical Advisory T5140.12, *Alternate Bridge Designs.* Washington, D.C.: Federal Highway Administration.

Transportation Research Record 1180. 1988. *Bridge Design and Testing.* Washington, D.C.: Transportation Research Board, National Research Council.

Transportation Research Record 1242. 1989. *Innovative Earth Retaining Systems.* Washington, D.C.: Transportation Research Board, National Research Council.

Troitsky, M. S. 1988. *Cable-Stayed Bridges An Approach to Modern Bridge Design.* New York: Van Nostrand Reinhold Company.

Troitsky, M. S. 1990. *Pre-stressed Steel Bridges Theory and Design.* New York: Van Nostrand Reinhold Company.

Turner, O. D., and Robert T. Reark. 1981. *NCHRP Synthesis 78, Value Engineering in Preconstruction and Construction.* Washington, D.C.: Transportation Research Board, National Research Council.

U.S. Department of Transportation. 1979. *FHWA Bridge Inspector's Training Manual.* Washington, D.C.: Federal Highway Administration.

Wahls, Harvey E. 1990. *NCHRP Synthesis 159, Design and Construction of Bridge Approaches.* Washington, D.C.: Transportation Research Board, National Research Council.

Weller, Clyde G., and John Crist. 1991. Timber bridges: background, attributes, national direction, and stressed timber. *Transportation Research Record 1291, Volume 1*:287–292.

# 13

# Intersections and Interchanges

Continuing where we left off back in Chapter 9, we come to the essence of redoing, improving, and reconstructing our highway infrastructure. Converting intersections to interchanges will be an assignment that highway architects are to experience for many years. Short of this is the challenge to improve at-grade intersections, which is where we shall begin.

## INTERSECTIONS

More conflicts, more capacity constraints, and more accidents take place at intersections than anywhere else in a highway network. Intersections are nodes in the system, and, as any medical practitioner can explain, nodes are trouble spots. Geometry and measures to accommodate driver expectancy are the means available to enhance efficiency of an at-grade intersection. It is mandatory to understand, however, that all these factors are interdependent and not mutually exclusive. They cannot be dealt with separately.

### Conflicts

Vehicles, pedestrians, bicyclists, and sometimes even railroad trains, all trying to occupy the same space, preferably not at the same time, single out the intersection as the highway component where the greatest potential for conflict occurs. As with ancient Gaul, all conflicts can be divided into three parts: crossing, diverging, and merging. The first of these, crossing conflicts, contributes to the most severe accidents and capacity constraints. The latter two, diverging and merging conflicts, create the rash of fender benders.

## Capacity Constraints

When a platoon of vehicles comes rumbling to a halt in front of a traffic light or a full stop sign, efficient passage of the procession is underachieved. In fact, even after coming to a stop and then proceeding, the constraints of left- and right-turning vehicles tend to slow the procession.

Operational aspects can also clog up the system. If an intersection area has many utility lines running under and over, every time some work or repairs are performed by utility personnel, the intersection gets into its disfunction mode.

The *Highway Capacity Manual* gets very technical in describing how to determine the capacity of an intersection. A highway architect probably only needs to know the very basics. So here are some simplified rules of thumb:

1. Hourly capacity of each through traffic lane in a signalized intersection can be estimated as 1,800 times green time divided by total cycle time. That is, if one third of the cycle time is green, 1800 × 1/3 = 600 autos per hour in each lane; if half the time is green, the total would be 900 autos per hour per through lane.
2. Minimum vehicle headway per lane is two seconds.
3. Unsignalized intersections require lots of guesswork to ascertain capacity. A developed approach is contained in *Transportation Research Circular 373*.
4. Each bus or single unit truck occupies the equivalent space of two or three autos.
5. Each multiple unit can be considered the equivalent of three to six passenger cars.
6. For free-turning movements, use 90 percent of the capacity of through lanes. This does not apply for non-free-turning movements.
7. Use these factors for rough estimate guidelines.

## Accidents

Multiple contributors to accidents crop up at intersections. Not only the conflicts, but all too often the measures to enhance capacity and safety come into play to do just the opposite of their intent. Overloading vehicle operators with too many tasks to accomplish in too short a time is one reprehensible practice. Traffic signals, prohibition signing, directional signing, and channelization, all making their appearance while the driver is playing dodge 'em might satisfy legal requirements, but do not necessarily contribute to a safe driving environment.

## Geometry

Highway architects have one weapon to wield in order to increase efficiency and enhance safety at intersection nodes: the layout of the intersection itself. Manipulating this weapon becomes extremely complex. Should a conscientious architect attempt to employ all the guidance offered in Chapter IX of the AASHTO *Green Book*, the results could resemble the proverbial camel created by a committee of equestrians. Very much like overloading vehicle operators, the *Green Book* guidance overloads the intersection designer with many mutually exclusive contradictions (e.g., adding pavement area, reducing pavement area, providing free movements, eliminating free movements, adding signs, removing signs, providing pedestrian refuges, reducing median widths). After a while, the highway architect begins to realize that the operative term is "trade-off" when dealing with roadway intersections. Perhaps the only factor that is consistent is to provide adequate sight distance in all directions. Yes, but what is *adequate* sight distance? Another geometric feature that is difficult to resolve is how much tracking of large vehicles (such as turnpike doubles and articulated buses) is to be designed into the turning paths.

Vertical alignment may also play a role in the selection of measures to accommodate the anticipated traffic. Steep grades on the approaches are real headaches; flat grades may lead to drainage problems should the intersection pavement area be extensive.

Accompanying sketches show how intersection designs in various parts of the country attack some of the more common difficulties. The reader might find it challenging to redesign or modify their geometries within the right-of-way constraints nominally evident and then determine if the redesign is an improvement or a worsening.

Right-turn movements in urban and suburban settings turn out to be a bigger bugaboo than the *Highway Capacity Manual* is willing to concede. If there is no free flow for right turns, approaching autos and trucks decelerating to make sharp 90-degree turns, sometimes waiting for pedestrians to cross, can slow the following vehicles down to unacceptable levels. Drivers following are tempted to pull out of the right lane and create havoc for traffic in the adjacent lane, whether it be moving in the same or opposite direction. Provision of free right-turn movements is a valuable tool for both safety and capacity. The illustration of the Church Street–Bernard Avenue intersection in Greeneville, Tennessee, contains free right turns for all movements (see Fig. 13-1). Note that in the three heaviest traffic right turns, the lane width has been increased from 12 to 18 feet. The stop bars (lines) are placed well back of the channeling to permit greater radii for the left turns. This does create some additional pavement area for vehicles to clear, something that has to be incorporated into the signal timing.

Cross roads having skew angles less than 60 degrees can have sight distance

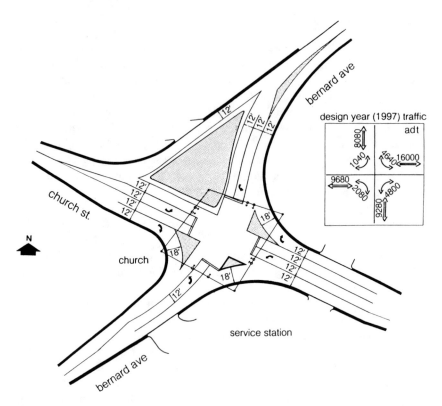

**FIGURE 13-1.** Church Street–Bernard Avenue, Greenville, Tennessee (Adapted from NCHRP Report 279).

problems. In fact, skewed intersections are asking for trouble, unless the predominant movements are the obtuse angle turns. As shown in the U.S. 41–Delaney Road intersection north of Chicago, considerable traffic comes from the north and turns to join southeastbound commuters in the morning (see Fig. 13-2). Then, during the afternoon peak, the same traffic returns from the southeast and turns right onto Delany Road. As one would expect at a signalized intersection, two left turn lanes are provided for the heavy movement, while only one (18-foot) lane is necessary for complementary free-flowing right turns of the same magnitude. (The right turns do not regularly have to wait for the traffic signal.)

A similar intersection, Piedmont/Blackland Road–Roswell Road in Atlanta has a skew of 45 degrees (see Fig. 13-3). Again, the south-to-southeast and northwest-to-north movements are predominant. Interestingly, two right-turn lanes are provided to complement the two left turns. Because of the skew, the sharp left and

Intersections and Interchanges   163

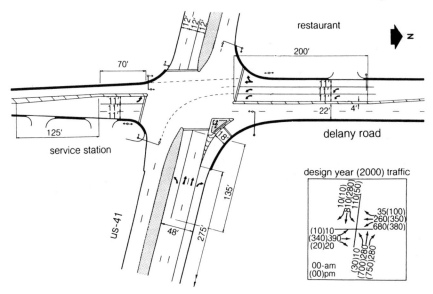

**FIGURE 13-2.** U.S. 41-Delany Road, Lake County, Illinois, (Adapted from NCHRP Report 279).

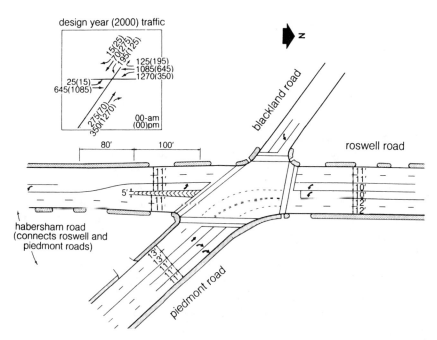

**FIGURE 13-3.** Piedmont/Blackland Road-Roswell Road, Atlanta, Georgia (Adapted from NCHRP Report 279).

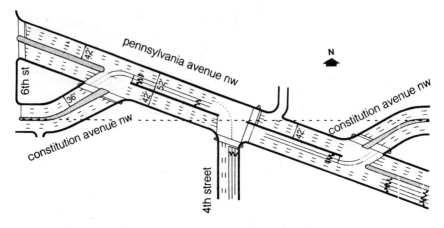

**FIGURE 13-4.** Pennsylvania Avenue and Constitution Avenue, Washington, D.C. (Adapted from NCHRP Report 279).

right turns from and to Piedmont are prohibited; Habersham Road provides a surrogate for these.

When the skew angle is found to be ridiculously small, a major realignment is in order such as that (a number of years ago) at Pennsylvania Avenue and Constitution Avenue in Washington, D.C. (see Fig. 13-4). The former skew angle was increased from 20 degrees (dashed line) to a more acceptable 50 degrees, and two new intersection locations were provided along Pennsylvania Avenue, which accommodates Constitution Avenue traffic for 800 feet. Widening Pennsylvania Avenue to nine lanes through this segment required reducing the median width. As a consequence, pedestrian refuge in the median is virtually nonexistent.

Provision for storage of vehicles waiting to enter an intersection is always chancy. The left-turn lane on 63rd Street eastbound on an intersection just south of Kansas City appears to have adequate storage, but the short lengths of turning lanes on Mastin Street between the frontage road and 63rd westbound cannot handle more than four autos, two in each lane (see Fig. 13-5).

When one or both intersecting streets are on a curve, superelevation and sight distance concerns mount. At the 1st Avenue North–Exposition Drive three-way intersection in Billings, Montana, there is the added complexity of significant left-turn movements from Exposition Drive (see Fig. 13-6). As in the U.S. 41–Delany Road intersection, two left turn lanes are complemented by a single right-turn lane. Note the staggered left-turn stop bars on 1st Avenue that allow for a flatter radius of turn from Exposition Drive.

Intersections and Interchanges    165

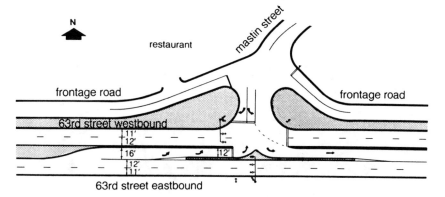

FIGURE 13-5. 63rd Street–Mastin Street, Merriam, Kansas (Adapted from NCHRP Report 279).

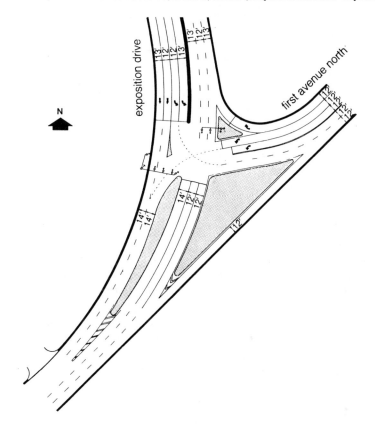

FIGURE 13-6. First Avenue North–Exposition Drive, Billings, Montana (Adapted from NCHRP Report 279).

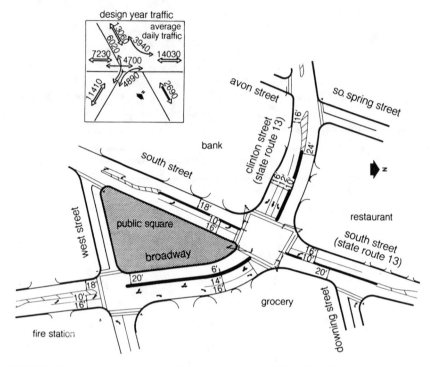

**FIGURE 13-7.** South Street–Broadway–Clinton Street, Concord, New Hampshire (Adapted from NCHRP Report 279).

In Concord, New Hampshire, the designers and traffic engineers seemed to be faced with an almost impossible problem at South Street–Broadway–Clinton Street (see Fig. 13-7). Curved alignment, close proximity of other streets, a multitude of access points to business properties, pedestrian activity, and a helter skelter street pattern all lead to clutter. Although the problems are not completely solved by channelization, the amount of traffic that travels through the locale has not been sufficient to warrant a great expenditure of funds. Many of the older communities in the northeast are faced with similar situations in which intersections function very poorly and extensive geometric improvements are desirable. Yet these are where the no-action alternative becomes very attractive.

## Driver Expectancy

Human factors enter into the mix to a great degree in the design and operation of an intersection. Here the architect has an edge over the engineer, presuming that the architect has achieved a balance in dealing with both people and things. Signs, pavement markings, placement of traffic lights, and, particularly, physical appearance of the site tend to dictate how vehicle operators, pedestrians, and bicyclists are likely to behave during their approaches to and excursions through intersections. The *Manual of Uniform Traffic Control Devices* (MUTCD) provides the basic technologic framework. After that, the architect must rely on those with backgrounds in the behavioral sciences. In 1989, the FHWA released a study that showed that motorists are not always inclined to obey traffic control devices. Adjustments are also necessary in places where elderly and tourist traffic is prevalent, as opposed to intersections habitually traversed by commuters two times a day. MUTCD standards do not always allow for such aberrations.

## HERESY

Before getting to the larger subject of interchanges, there are two topics where an iconoclastic highway architect will encounter a great deal of opposition. One is any suggestion that reducing the number of lanes can enhance capacity; the other is recommending that an existing intersection be replaced by a grade separation, with no ramps. (The AASHTO *Green Book* mentions this alternative, but does not dwell on it for very long.) Such recusant doctrines apply only in rare instances, but can be valid as in the following two somewhat hypothetical citations derived from actual cases.

Main Avenue is a six-lane divided urban arterial (see Fig. 13-8). At Third Street there are many turning movements that back up Main Avenue traffic so that the approaching lanes become storage during much of the day. (One city councilperson facetiously suggested installing parking meters on all the lanes.) Obviously, the problem is at the intersection and, no matter how many lanes on Main Avenue, will continue to be congested. Thus, the first order of business is to clean up the bottleneck at Third and Main. Property values are high, so major right-of-way takings are not feasible.

Actions to address the problem include providing a free right turn lane at the intersection, reducing Main Avenue from a six-to a four-lane divided facility and installing advance signing. What would be accomplished?

1. Vehicles approaching to make a left turn at Third Street will be directed to the left lane.
2. Through and right-turn vehicles will be directed to the right lane so that, when the free right-turn lane becomes available, right-turning traffic moves out of the way, no longer to obstruct or conflict with through movements.
3. Twenty-two feet of pavement width is to be removed from maintenance responsibility; savings made here can be allocated elsewhere.

168   Highways: An Architectural Approach

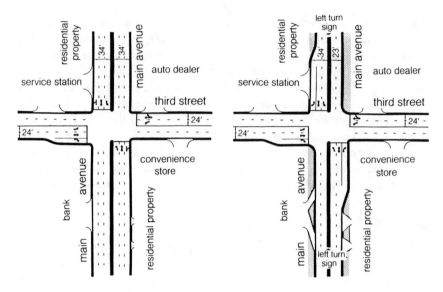

**FIGURE 13-8.** Main Avenue–Third Street: Before and after suggested modification.

4. Traffic noise and air quality degradation will have been moved 11 feet farther from businesses and residents facing Main Avenue.
5. Capacity of the two approaching lanes will still exceed the maximum green time flow rate possible at the intersection.
6. Stop bars on Third Street can be moved closer together to reduce intersection clearance time.
7. Increased radii for right-turn movements from Third Street are possible.

Essentially, the intersection will have been made more efficient. Levels of service on both intersecting streets can be improved to a small extent. Do these meliorations appear to be minute? At first, yes, but let's delve into one issue further.

Traffic engineers, in sort of a knee jerk reaction, are tempted to widen intersections whenever they get clogged (the intersections, not the traffic engineers). There is justification for doing so in most cases, but not always. Should the width of an intersection be reduced (in this case by 11 feet), pedestrian crossing time is likewise reduced (in this case by four seconds). Four seconds translates into a potential capacity increase of eight vehicles per cycle for four lanes, which, in turn, means an additional 200 vehicles per hour on Main Avenue, if the minimum signal timing is governed by pedestrian crossing time.

We can demonstrate other instances where reduction in the amount of pavement can contribute a salutary effect on traffic movement. The reader is urged to ponder this unorthodoxy.

Another paradoxical case can be cited at a growing suburban intersection. Traffic at Pine Boulevard and State Route 17 continues to increase with continued development along both arterials (see Fig. 13-9). Both routes have been widened to handle considerable through traffic. However, continual street and intersection widening appear to do little to reduce rush hour congestion, and adjacent property values are escalating.

*Solution*: Separate Pine and Route 17 by running Pine over Route 17 on a bridge structure (a flyover). There is not enough room within existing right-of-way to allow for interchange ramps. What would be accomplished?

1. Free flow on both Pine and Route 17 would eliminate the bottleneck and permit, at least in this location, both arterials to operate at a comfortable level of service.

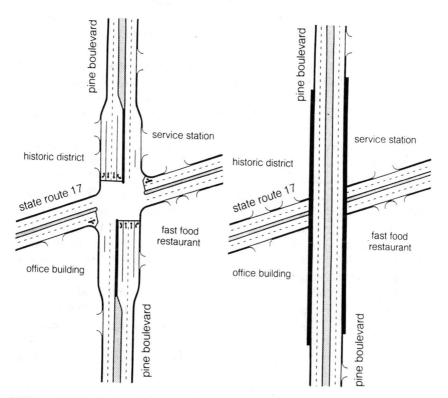

**FIGURE 13-9.** Pine Boulevard–State Route 17: Existing intersection and suggested overpass.

2. Turning movements will have been eliminated, but inter-access between Pine and 17 can be accomplished at other locations in the street network.
3. Nearby merchants will be madder'n hell until they realize that, by modifying their own access to the arterials, business could well improve.

These two scenarios are not exactly typical, but they are presented to encourage the highway architect to think through the myriad possibilities and alternatives worthy of consideration when an intersection problem becomes especially knotty.

## INTERCHANGES

Chapter X of the AASHTO *Green Book* shows seven interchange types. There are variations within each type, but the highway architect is not likely to be required to design many of the fancier ones; they've already been designed and put into operation. The future will see modifications to existing interchanges and conversion of intersections to interchanges. Yes, even the other way around: where existing interchanges do not function well or where they are no longer needed, they may be removed.

Interchanges are costly when only the first costs (design and construction) are taken into account. When user benefits resulting from reduced congestion are balanced against user costs and thrown into the financing mix, converting clogged intersections to the more free-flowing interchanges can come out smelling pretty good. Planners have probably done a cost analysis well before a highway architect is directed to convert an intersection to an interchange. Still and all, it does no harm to review the cost analysis before proceeding with the design. If no cost analysis has been undertaken for a specific site, the architect ought to put together, as a minimum, tabulations relating traffic volumes, speeds, travel distances, delays, and other items involved in user costs against construction costs. If the results clearly favor building an interchange, go to it. If not, a further analysis should be pursued that needs to include present worth, maintenance costs, life cycles of the various elements, and the anticipated safety benefits: a complete economic analysis.

The highway architect will usually be responsible to determine which type of interchange would be most beneficial within the traffic/financial/right-of-way constraints at any one location. Type selection, however, becomes increasingly limited when contrasted to the opportunities engineers had in the past when designing, rather than redesigning, highways. (In those days past, retrofitting played an insignificant role.) Directional and cloverleaf interchanges require either oddball movements or too much right-of-way; thus, we had better concentrate on the traditional diamond and its mutations.

## Diamonds: CDIs, TUDIs, SPUIs, and Stacked

Most common in both rural and urban settings, the diamond interchange has limitations in its adaptability. Too much traffic on the minor road or many turning movement demands cause gridlock. When this happens, it may be necessary to look more broadly at the roadway network, rather than expending all one's effort at a single node.

A simple diamond has ramps that are fairly well spread apart so as to create two separate intersections on the minor crossroad. This works well in rural locations where a wide right-of-way is normally available. In the more congested areas, the compressed diamond interchange (CDI) comes into play. The ramps are brought closer together—200–300 feet on each side of the grade separation. But then we create a potential problem: intersections closer together than desirable with the spectre of queues of autos and trucks backing into one of the intersections while waiting at the traffic signal in the other intersection. Often, there are frontage or service roads close by that further exacerbate congestion.

Next we come to the tight urban diamond interchange (TUDI), where the two ramps are brought as close together as possible and the signals interconnected so that they are able to prevent gridlock on the minor crossroad (see Fig. 13-10). But then what happens if traffic coming off the major crossroad has to stop for lengthy periods and gets backed up onto the traveled way? Although not restricted to the TUDI, adequacy of the off ramps must be handled in the design phase, rather than calling it an operational problem. *NCHRP Synthesis 35* delves into this issue. Also, there is the matter of high left-turn movements from the minor to the major crossroad putting additional strain on TUDI operation.

One rather recent adaptation of the diamond is not described in the 1990 AASHTO *Green Book*: the single point urban interchange (SPUI) (see Fig. 13-11). First appearing in Florida and in western Illinois in the early 1970s, the SPUI places the two intersection movements of the TUDI at a single point under or over the major crossroad (see Fig. 13-12). The single point operates with traffic movements equivalent to those of a large intersection. Although not a cure-all, the SPUI has had reasonable success where there is lots of interchanging (left turns) between the two crossroads.

Next in the quest for optimum efficiency within minimum right-of-way is the stacked diamond (see Fig. 13-13). More grandiose than the SPUI, a stacked diamond separates the single point intersection from both crossroads, creating in the process three, rather than two, levels. Additional structure and associated earthwork raise costs to sometimes unacceptable figures in spite of the increase in efficiency offered by the stacked diamond.

A review of the abundant literature on the subject, beginning with the AASHTO *Green Book*, can provide the highway architect considerable opportunity to become sated with information on potential interchange designs and their hoped-for capabilities.

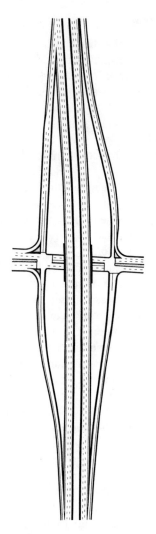

**FIGURE 13-10.** Tight urban diamond interchange (TUDI).

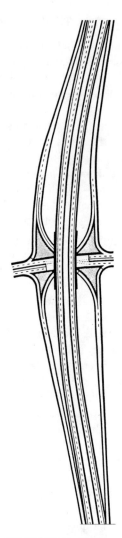

**FIGURE 13-11.** Single point urban interchange (SPUI).

172

Intersections and Interchanges    173

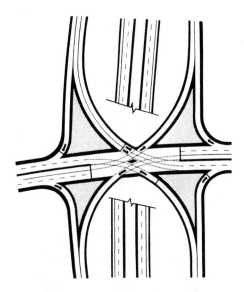

**FIGURE 13-12.** Cut-away view of the SPUI showing turning movements at the single point.

## Pedestrians and Bicyclists

Crosswalks are common at intersections. Sensible bicyclists will, at busy intersections, dismount and become temporary pedestrians. Sometimes pedestrian overpasses are constructed at high-hazard locations, but they tend not to be used unless foot traffic is prevented from crossing at-grade by some type of barrier.

Providing for pedestrian movements at interchanges is not so simple; providing for safe and sane bicycle movements at interchanges can get to be ridiculous. Diamond interchanges, with their myriad turning movements and signal interphasing, can be reduced to complete ineffectivity if pedestrian phases are to be incorporated. And worse, SPUIs, with traffic coming to and from every which way at any time, can panic even the hardiest pedestrian. The best rule is probably to divert pedestrian (and bicycle) movements upstream and downstream from any interchange that contains traffic signals.

## References

AASHTO. *A Policy on Geometric Design of Highways and Streets (Green Book).* 1990. Washington, D.C.: American Association of State Highway and Transportation Officials.

Anderson, Dudley G., *et al.* 1977. *A Manual on User Benefit Analysis of Highway and Bus-Transit Improvements.* Washington, D.C.: American Association of State Highway and Transportation Officials.

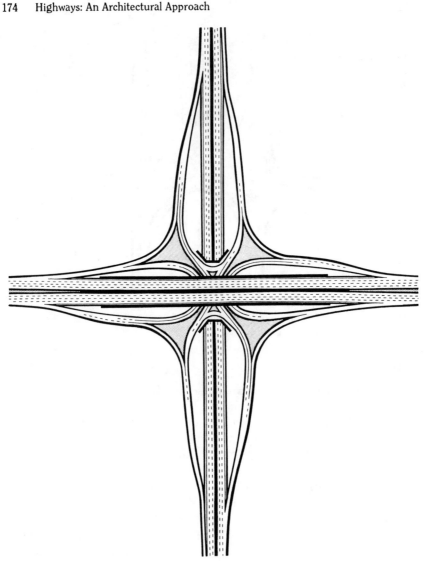

**FIGURE 13-13.** Stacked diamond interchange.

Batchelder, J. H., *et al.* 1983. *NCHRP Report 263, Simplified Procedures for Evaluating Low-Cost TSM Projects*. Washington, D.C.: Transportation Research Board, National Research Council.

Bowman, B. L., J. J. Fruin, and C. V. Zegeer. 1989. *Handbook on Planning, Design, and

*Maintenance of Pedestrian Facilities*, FHWA-IP-88-019. McLean, Virginia: Federal Highway Administration.

Hauer, Ezra. 1988. The safety of older persons at intersections. In *Special Report 218, Transportation in an Aging Society*, pp. 194-252. Washington, D.C.: Transportation Research Board, National Research Council.

*Highway Capacity Manual, Special Report 209*. 1985. Washington, D.C.: Transportation Research Board, National Research Council.

*Manual on Uniform Traffic Control Devices* (MUTCD). Regularly updated. Washington, D.C.: Federal Highway Administration.

Mason, J. M., Jr., et al. 1986. *NCHRP Report 283, Training Aid for Applying NCHRP Report 263*. Washington, D.C.: Transportation Research Board, National Research Council.

NCHRP Synthesis 35. 1976. *Design and Control of Freeway Off-Ramp Terminals*. Washington, D.C.: Transportation Research Board, National Research Council.

Neuman, Timothy R. 1985. *NCHRP Report 279, Intersection Channelization Design Guide*. Washington, D.C.: Transportation Research Board, National Research Council.

Pietrucha, Martin T., et al. 1989. *Motorist Compliance With Standard Traffic Control Devices*, Report RD-89-103. Washington, D.C.: Federal Highway Administration.

Transportation Research Circular 373. 1991. *Interim Materials on Unsignalized Intersection Capacity*. Washington, D.C.: Transportation Research Board, National Research Council.

Transportation Research Record 1114. 1987. *Traffic Control Devices and Rail-Highway Grade Crossings*. Washington, D.C.: Transportation Research Board, National Research Council.

Transportation Research Record 1141. 1987. *Pedestrian and Bicycle Planning with Safety Considerations*. Washington, D.C.: Transportation Research Board, National Research Council.

# 14

# Roadside

Nothing in the highway business cries out more for an architectural approach than does the roadside. Environment, life-style, and coordination with adjacent communities are all found here. Determining how a road fits in with its surroundings—and how the surroundings fit in with the road—can be looked at as a rewarding challenge for the enterprising highway architect. Eight of the more prominent aspects of the roadside, in no particularly logical sequence, are sketched in this chapter.

## VEGETATION

Highway architects with a background in horticulture or landscaping are apt to spend too much time and effort in revegetating the roadside, to the detriment of other tasks. This is not to be taken as disparaging; most persons involved in highway development or redevelopment spend too little time with this matter. (There should be a balance somewhere.)

### Macro-scale—The Broad Picture

Except in the most arid climates, planting of vegetation is the most suitable means to cover barren ground—both in medians and within highway right-of-way outside the roadway prism. As pointed out in the chapter on drainage, vegetation controls erosion; it also contributes to pleasing aesthetics.

Soil conditions and climate dictate choice and arrangement of plant materials. Urban locales usually require considerably more (and larger scale) plantings than comparable rural locations. In the latter, extension of native growth often does quite well, while in the former it is essential that the plantings be utilitarian. Consequently, the planting scheme should begin with an overview before the specifics are worked out. Coordination with the total environment takes priority over an architect's pet prejudices regarding favorite plants and trees.

We are dealing with highways; thus, the motorists and safety issues must predominate. From the viewpoint of the motorist, scene changes and roadside tableaus ought to be commensurate with vehicle speed. Tree plantings ought not to become hazards. Sight distance standards have to be observed.

## Micro-scale—The Specifics

The simplest way to find out what plant species will do best is to visit the site and observe and catalog the predominant vegetative elements. Short of doing this would be to do an investigation of plants that grow well in comparable soils, topography, and climate. In matching plants to climates and topography, site characteristics may play an important role. Slopes facing to the south and west (in the northern hemisphere) receive greater exposure to sunlight than flat surfaces and slopes facing to the north and east. This is why stretches of highway in cuts often show different vegetation patterns on opposite sides of the road.

Legumes and grasses perform a somewhat different roadside function than bushes and trees do—parallel and corresponding to the different roadway functions performed respectively by pavements and bridges. One is for ground cover, the other for elevation and separation. Another way of stating it is that turf is two-dimensional; bushes and trees, when placed in clusters, are three-dimensional. (Widely spaced, individual trees such as along the right-of-way line as favored by

**FIGURE 14-1.** Interesting roadside management: extensive tall vegetation on the north side of the road; barren on the south side.

some highway agencies, are pretty much nondimensional. They may remind passing motorists of soldiers guarding the ramparts against an imminent attack by the heathen hordes.)

### Restoration

Examples of roadside restoration directed at adaptation may be reviewed in the *TR Record 913* summaries of research done in Wyoming (pp. 4-10), Massachusetts (pp. 11-14), and New Jersey (pp. 23-28). Not to be overlooked is the requirement that plants be resistant to atmospheric pollutants that occur in urbanized areas and resistant to salts in rural and suburban areas that are subject to snow-removing operations. There was also a directive contained in the Surface Transportation and Uniform Relocation Assistance Act of 1987 that native wildflowers are to be part of highway landscaping projects sponsored (in part) by federal dollars.

Coloring the highway landscape with flowers, particularly when the government directs one to do it, ought to be one of the more pleasurable aspects of designing and building highways. Texture and color are to match up throughout the seasonal changes. Information that may help the architect in flower selection can be obtained from the The National Wildflower Research Center Clearing House in Austin, Texas, or from a local branch of the Federation of Garden Clubs.

### Screening

Hiding some of the more unsightly roadside blemishes (i.e., auto graveyards, garbage dumps) is something that roadside plantings do best. They do have limitations, though. Unless at least 100-feet thick, vegetation does little to reduce noise. However, modest height plantings have been used very effectively, primarily in medians, to screen headlight glare from approaching vehicles (see Fig. 14-2).

### Fire Hazards

Sometimes, in the arid climates, roadside plantings look good in the springtime and acceptable through early summer. Late summer and autumn may turn such plantings into fire hazards—something else for the highway architect to be on the lookout for.

### Snow

Will the proposed planting scheme create hazardous snow drifts? Are plants going to make it difficult for maintenance personnel to push snow off the road? On the other side of the coin, plantings can be useful as living snow fences at strategic locations in order to prevent snow from drifting onto the highway (see Fig. 14-3).

**FIGURE 14-2.** Median plantings can provide an attractive headlight screen.

## Maintenance

A prevailing problem with roadside vegetation has been poor (or improper) maintenance after a highway segment has been completed and opened to traffic. For this reason, it is incumbent on the architect to be aware of just what future maintenance effort will be available for any vegetation or revegetation items to be contained in a construction contract. (Examples of certain maintenance techniques used in Maryland, Alabama, Idaho, and Indiana are reviewed in *TR Record 1075*, pages 1-18.) Fundamentally, there are two dictates that must be met:

- Growth and survival of plant materials are directly related to the timing and amount of water applied, primarily during the first growing season. Both overwatering and underwatering must be avoided.

**FIGURE 14-3.** Where adequate moisture is available in the colder climates, a living snow fence can effectively keep snowdrifts from obstructing highway traffic at critical points.

- Plants selected must be able to adapt quickly to the highway environment and successfully compete with other vegetation.

**Successful Roadside Revegetation**

Highway roadside seeding and plantings are more extensive than one usually encounters when landscaping a quarter-acre home plot. Wide stretches of grass seeding require tackifiers to hold the soil together in certain climes.

Consultation with a nursery operator or a landscape architect is occasionally a good practice where roadside revegetation may go beyond the typical topsoil and seeding regimen included in many highway contracts. Considerable information on revegetation species for all parts of the country can be gleaned from *FHWA/TS-82-212*, pages 52-60 and 81-93.

Planting plans may well be presented on drawings that are separate from the roadway plans, although they should be tied to the roadway plans by baseline or centerline stationing. On the planting plans, adjacent land uses, topography, utilities, roadway construction limits, slopes, and similar features should be identified. On less complicated projects, planting plans can appear on the same sheets as the construction plans. However, in either case, the plan sheets need to show clearly the planting schedule (list).

# WILDLIFE

Periodically, newspapers fill up some remnant spaces with features on how various highway departments have provided road crossings for turtles, salamanders, ducks, and other precious members of the animal kingdom. Actually, the issue is more related to safety if one can imagine the skidding potential when a vehicle drives over several thousand migrating salamanders. Most roadside wildlife conflicts concern ungulates (antelope, deer, elk, moose, and other hoofed creatures.)

Much of the following comes from results of a 1982 survey made by the Federal Highway Administration, Central Direct Federal Division.

**Visibility**

Providing broad visibility for both motorist and beast by clearing roadsides of most or all vegetation higher than 6 inches has been one means used by some highway departments to reduce road kills and vehicular-wildlife collisions. Yet wildlife biologists appear to have divided opinions on how cleared roadsides affect wildlife. Differences of opinion may be attributed to uniqueness of each road section, changing wildlife population sizes and needs, changing habitat characteristics, shifts of crossing corridors, variations in local topography, fencing design, revegetation, changes in traffic volume, and vehicle speeds. All these factors can combine to produce a complex picture in any road crossing/roadkill situation. The prevailing opinion is that roadside clearing does help to reduce road kills because cleared roadsides offer better sight distance and reduce roadside blindspots, thereby providing motorists with more reaction time to observe crossing wildlife and to brake or otherwise avoid potential collisions. Naturalists and some environmentalists extend a caution in this regard, however. Overly extensive clearing of the right-of-way may not only reduce the value of habitat and crossing corridors within the right-of-way but may also reduce the value and carrying capacity of habitat adjacent to the right-of-way due to loss of screening cover between the road and this habitat. In some areas where extensive land clearing has occurred, roadways may be bordered by older and denser growth that may comprise the only remaining cover habitat in the area. The advantage of roadside clearing to reduce roadkill potential may have to be traded against overly extensive clearing, which can reduce habitat values and wildlife population sizes.

## Traffic

Highway safety analysts point out that increases in traffic volumes and/or vehicle speeds result in corresponding increases in roadkills of crossing wildlife. Paradoxically, traffic volumes can increase to a point at which the vehicles become a moving barrier to crossing wildlife; roadkills then decrease due to fewer crossings. It is also widely recognized that vehicle speed is related to motorist perception of the safety of the road and that roadside clearing is one of the factors contributing to a safer feeling, higher speeds, and, possibly, reduced driver alertness, leading to more severe vehicle-wildlife altercations. In proportion to their numbers, heavy trucks cause the greatest number of roadkills, since these vehicles often travel at high

**FIGURE 14-4.** Deer kills on the highway are a widespread problem.

speeds and may sustain relatively little damage from collisions with wildlife, while drivers may not be able (or desire) to take evasive action.

**Right-of-Way**

The highway architect, when confronted with a rural or suburban roadside, must realize that animals (primarily cattle and other ungulates) are very often attracted to revegetated road rights-of-way to feed on nutritious grasses that are planted for erosion control or grow up naturally over time. These grasses often "green up" earlier in the spring than surrounding areas because the roadside is more exposed to sunlight, moisture, and other microenvironmental conditions that cause snow cover to clear and growing conditions to occur sooner. Likewise, right-of-way grasses continue to grow and remain free of snow cover and leaf litter longer into autumn when other forage areas may be less attractive or less available. In spring and autumn, therefore, a marked concentration of wildlife in grassy rights-of-way is to be expected. This effect is most striking in forested sites with few other openings and where other locations have been overgrazed by domestic livestock. The problem of wildlife attraction to grassy rights-of-way in autumn and spring is compounded by the fact that these are also periods of high seasonal wildlife movement. Wildlife feeding near the road significantly increases the potential for road crossings and roadkills as well as the opportunity for poaching or harassment. One possible measure to assauge these problems is to revegetate roadsides with unattractive/unpalatable grasses to avoid drawing wildlife into the right-of-way corridor to forage in a long, narrow, hazardous pasture. Nevertheless, over time, natural plant succession in undisturbed rights-of-way may reestablish wildlife-attracting grasses, forbs, and shrubs. Accumulated salt from winter maintenance (where salt is applied) may also attract wildlife in spring, particularly if other salt licks are not readily available.

All of this brings up the matter of right-of-way fencing, a complicated tangle for the highway architect faced with satisfying sometimes mutually exclusive factors such as wildlife concerns, motorist safety, efficient transportation, and roadside aesthetics. High (7 to 8 feet) restrictive fencing, if properly installed and maintained, is a most effective means of reducing roadkills, although such fencing is costly and infrequently installed because it can be so complete a barrier that, unless some crossing mechanism is provided, wildlife can be (and have been) excluded from valuable and required habitat and migration corridors.

Most wildlife biologists believe that the more typical right-of-way fencing (4 to 5 feet high) does not significantly restrict wildlife movement (except for antelope among the ungulates) and has little significance in reducing road crossings and consequent roadkills. Placement of typical fencing in grassy rights-of-way may influence the potential for roadkills, since deer will readily jump the fence to graze the right-of-way (should grazing acreage be limited). This can put large numbers of deer close to the road, away from protective cover and with little deterrent to crossing the road. Fencing (along with one-way deer gates) under these circumstances ought to be installed so that at least 25 yards of grazing area remains behind the fence (i.e., between the fence and edge of cover). Deer then may graze behind the fence, and the fence would become a more practical deterrent to road crossing. Of course, there are certain disadvantages to placing the fence within the right-of-way instead of on the right-of-way line. However, if roadkills are a significant problem because of right-of-way grazing, consideration should be given to fence location. If both roadsides are revegetated with unattractive/unpalatable grasses (as mentioned earlier), the importance of fence location may be minimized.

**Wildlife/Motorist Interactions**

Actually, we'd like to avoid interactions, so the architect ought to be alerted to some common circumstances:

- Much wildlife activity follows both seasonal and daily patterns; in temperate climates, activity occurs in spring and fall; most daily activity occurs from dusk to just after dawn.
- Ungulate activity (particularly for deer and elk) is strongly seasonal, with movement to and from winter and summer ranges in spring and fall and relative inactivity in summer and winter. Spring is a peak movement period, due to the need for forage following depletion of fat reserves, dispersal of yearlings, calving, and movement to summer ranges. Autumn is a peak movement period because of the need for forage to build fat reserves, rutting (and mating) season, hunting seasons, and movement to winter ranges.
- Ungulates move daily (often several times at night) between cover/resting sites (usually within forest or shrub cover) and feeding/watering sites.
- Larger wildlife populations mean more wildlife road crossings and more roadkill potential. Roads through important wildlife areas need to be signed for crossing wildlife potential.
- It appears that permanent human activities and structures result in reduced wildlife populations and lessen roadkill potential. More human activity increases the tendency of wildlife to be active at night.
- One animal seen crossing (or in the right-of-way) increases the probability that more may be nearby or following; eyeshine may be the only indictor of wildlife presence at night.
- Wildlife will often bolt if startled by vehicle headlights and may bolt in any direction; on the other hand, they may disregard or seem unaware of the hazard of an oncoming vehicle until collision is unavoidable. Speed reduction, particularly from dusk to dawn and upon sighting wildlife, is the most effective means for drivers to avoid collisions with wildlife.

It may be noted at this point that animals oftentimes have more common sense than many drivers. For this reason there has been some success attributed to informing not only the drivers but also the animals of the dangers at road crossings. Wildlife will often cross at predictable and established locations (e.g., game trails (especially in winter), locations with sufficient cover, along topographic features such as the transition between hills and valleys or cuts and fills, along stream bottoms). So far, "wildlife signing" has taken two forms:

1. High frequency sound warnings that activate when vehicles approach major game crossings.
2. Triangular mirrors that reflect vehicle headlights into the likely wildlife paths and trails. (*FHWA/RD-82/061* concluded that "there is no statistically valid evidence that...reflectors reduce vehicle-deer collisions." This nonconclusion has been further developed in the debate contained on pages 35-43 of *TR Record 1075*. Additional research is still going on.)

Much has been written concerning wildlife-vehicle confrontations. Very little has yet been directed toward other wildlife interests. Something like roadside plantings to attract songbirds (*TR Record 1075,* pages 19-20) may be a coming trend. A discussion regarding not only songbirds, but also amphibians, reptiles, and small mammals shows up in *FHWA/TS-82-212*. Pages 2 and 4 of the document lead the reader to topics of specific interest.

## UTILITY RELOCATIONS

Why is it that, invariably, within six months after a new pavement surface is placed, some utility company comes in and digs up the pavement to put in another pipe or conduit?

(No response.)

Can't the highway engineers arrange to have all this utility work done before the pavement is put down?

(Still no response.)

Is it because people in the highway department won't communicate with their counterparts in the waterworks? In the phone company? In the sewer department?

(And still no response.)

Should an architect opt for coordination, he or she might refer to some liaison procedures discussed on pages 7-9 of *NCHRP Synthesis 115*.

### Underground Utilities

Water mains, gas mains, laterals, sewer lines, electric conduits, pipelines, and assorted paraphernalia are always found underneath urban and suburban highways—sometimes on purpose, sometimes by accident. An interesting, though hazardous, recreational pursuit in a densely populated urban neighborhood is to watch a road construction crew remodeling an existing street. Each time the backhoe chomps up a half-yard or so of soil, the operator and inspectors look to see if any unlogged utility lines have been discovered/disconnected/dismantled/destroyed. When a 50-foot high geyser suddenly sprouts, a dozen workers run about looking for water main valves to turn off. (In some communities, certain valves operate clockwise and others counterclockwise so that the worker may never be sure whether the valve is being turned off or turned on.) When a bucket pulls up a 24,000-volt electric line, the designer or architect, complacently seated at an office desk (or computer terminal) some miles away; is blissfully unaware of all the turmoil caused by omissions or incorrect spotting on the plans of underground utilities.

### *Longitudinal Lines*

Those underground utility lines that will parallel the roadway are usually now placed in a utility corridor located away from under the pavement, sometimes in the shoulder subgrade, under sidewalks, or near the right-of-way fence line. The architect or roadway designer is responsible for the hierarchy, space allocation, and appropriate depths to be specified for each line. This makes it mandatory for the architect to consult with each utility company/agency to determine and/or ascertain their requirements—both in the present and for the foreseeable future.

### *Transverse Lines*

These are the biggest headaches because they become the most obtrusive if not planned for properly. Opening a just-completed pavement to put in a lateral line leaves a scar across the travelled way that is almost impossible to repair adequately. Larger transverse lines are often sleeved or encased. (Appendix A of *NCHRP Program Report 309* provides guidelines for protecting pipelines through highway roadbeds.) But whatever is required to provide necessary utility services to both sides of the road is ultimately the responsibility of the highway architect. And most utility companies/agencies and their personnel are most cooperative, accommodating, and helpful. It also makes for a much cleaner job if existing utility installations can be bird-dogged and accurately shown on the plans, rather than leaving it to the construction contractor in the field to find them by trial and error.

### Aboveground Utilities

Power/telephone poles and fire hydrants are the most common utility installations out in plain sight. Occasionally suggestions are made that power and phone lines could, for aesthetic and safety reasons, be placed underground. (*TR Record 970*, pages 52-64, contains a number of tables, along with text, that can be used to make some judgments as to the economics and propriety of going underground.) Again, in dealing with aboveground utilities, the same dictum applies as with underground utilities: the highway architect has the responsibility to make the necessary contacts and oversee the coordination of utility relocations generated by the highway design. Most major highway agencies have manuals for dealing with potential utility conflicts and policies for entering into agreements with utility and railroad companies.

### Railroad Crossings, Overpasses, and Underpasses

Although not exactly a roadside utility, railroads require the same type of coordination and agreements as do utility companies. Motorist perceptions at railroad grade crossings are discussed on pages 52-59 of *TRR 1160*.

## ACCESS

Along with highway operation are interests of the abutting landowners who have rights-of-access (consistent with their needs and the safe and efficient operation of the highway). Road users, of course, have rights to freedom of movement, safety, and efficient expenditure of public highway funds. It is generally considered to be the highway agency's responsibility to regulate and control the location, design, construction, and operation of access driveways and to reconcile, and, to the extent feasible, to satisfy the needs and rights of both the road user and abutting landowners. When thin strips of property are to be taken for highway right-of-way, there still must remain an adequate buffer area between the travelled way and roadside objects and edifices. This may require some regrading, cutting, or filling to insure ample sight distance for traffic operation, proper drainage, suitable slopes for maintenance operations, and good appearance. Trees, shrubs, ground cover, or other landscape features existing initially may have to be removed or adjusted. The buffer area must be free of any encroachment that would hinder traffic. All driveways and buffer areas ought to be constructed so as not to impair drainage within the highway right-of-way or alter the stability of the roadway subgrade. A particular sore spot in recent years involves quality of drainage water (originating at service stations or special industrial processing plants) that, after passing through highway drainage systems, is directed into irrigation canals.

Once the architect has pondered the various philosophical points, the bottom line says that he or she must abide by the regulations in force for the control and protection of the responsible highway agency's right-of-way.

# NOISE ABATEMENT

Touched upon in Chapter 8, the problem of highway noise and how best to reduce its impact is an issue that, even if it is ignored, won't go away. Vehicles create noise from:

- Exhaust pipes and stacks;
- Fans;
- Engines; and
- Tires against the pavement.

Since each of these sources generates noise within a different range of sound frequencies and from a different height above the ground, they can be attenuated with varying prescriptions.

## Noise Sources

Large trucks and semis are usually diesel powered; they are always loud. Their noise generation is somewhat independent of the speed at which they operate. From any given receptor location, the lower the truck speed, the greater the duration of noise. Passenger cars, on the other hand, tend to be noisier as their speed increases. A steady traffic flow at high speeds is noisier than the same flow at low speeds.

Suppose we look at a typical situation where the volume is 1,000 vehicles per hour with 6-percent heavy trucks. Presuming a level of service (LOS) C is prevalent, the noise level would probably be at its highest. Autos will be moving along at a good pace and a truck passing by every minute or so. A higher level of service (A or B) implies that the traffic volume is lower; a lower level of service (D or E) implies that traffic is moving more slowly. In both these latter cases, the noise levels (emanating from LOS A, B, D, and E) are lower than at LOS C—an interesting paradox.

Although we might expect the highest decibel readings at LOS C, it isn't always true. Sometimes high-speed expressway commuter traffic at LOS B or lower speed high-truck-percentage, horn-honking LOS D or E traffic produces greater noise nuisance.

## Attenuation Measures

Reducing the noise impact on nearby receptors is most often accomplished by either depressing the highway profile below the surroundings, providing solid noise barriers, or both. Trees, shrubbery, and other non-opaque measures, although inducing a psychological effect, do very little to reduce the decibel level.

Possibly the most cost-effective means of reducing traffic noise a small amount (where second story and higher receptors are absent) is to create a 2–3-foot high berm or mound parallel to the roadway. Tire noise will be reduced, and some of the

188  Highways: An Architectural Approach

fan and engine noise abated. Depressing the roadway profile 2–3 feet will also accomplish the same end. However, if major noise abatement is requisite, the height of berm/wall or depression of the grade line may have to be increased 12 feet or more. And this action would not be effective where receptors are more than a few feet above ground level.

Overhead viaducts are particularly troublesome because low-frequency noise conducted down through the structure compounds the higher frequency noise convected from traffic on the structure.

Community involvement, if prudently handled, can be a suitable means to determine the type and extent of noise attenuation measures that may be acceptable at specific locations. Judgment, together with local, state, and federal regulations, determines the extent of noise suppression required. Where many sensitive receptors are nearby, reduction of highway-generated noise is mandatory. However, "sensitive receptors" do not include foundries, rolling mills, and rocket testing facilities.

## REST AREAS

Several broad-base suggestions were put forth in Chapter 3 (in connection with multiple use and joint development) for rest area facilities. In this chapter we shall restrict ourselves to those rest areas that are fully the responsibility of the highway agency. Inasmuch as each highway agency has its own ideas on rest areas, the policies and standards of the sponsoring agency will override any recommendations or suggestions contained herein.

**FIGURE 14-5. & 14-6.** Noise walls are simpler to erect in sections;... ...however, unless they are butted tightly together, considerable noise goes through the openings and they lose much of their usefulness.

## Hierarchy

Rest areas come in various sizes, shapes, and denominations. From the lowest to highest in eminence, we have:

- *View sites*, which may be nothing more than places for motorists to pull off the road, stash some refuse, and gaze at the wonders of nature.
- *Truck rests* for teamsters to take a quick doze, inspect cargo, and check brakes should a long downhill stretch be forthcoming (implying that there may be some runaway truck ramps placed strategically along the way as complements to the truck rests).
- *Ports of entry* at or near state lines.
- *Rest stops* for travelers, the most common type of rest area.
- *Welcome centers*, which provide staff attendants, information, and other amenities.

## Spacing and Locating

All of the above are located at suitable sites to serve their stated functions. In addition, rest stops are usually spaced about an hour driving time on heavily traveled routes. (Rarely are rest stops provided on roads where the ADT is less than 1,000.) For the three higher types, availability of water and electricity is essential. Sewage disposal must also be considered.

## Size

View sites might only be large enough to accommodate one auto and a barrel for litter disposal; at scenic overlooks they may be considerably larger, expecting someday to be upgraded to rest stops. Truck rests and ports of entry are usually restricted to the area necessary to function effectively.

Rest stops throughout the country average at least 20 acres, with 5 to 10 of those acres developed (pavement, structures, walks, and picnic tables). The remaining property is used as buffer, landscaping, pet exercise, or reserved for future expansion. Some full-fledged rest stops are provided on only 4 acres of land; others occupy nearly 100 acres. Welcome centers are about the same size as rest stops, but more extensively developed.

## Facilities and Services

Parking and refuse barrels are all that one may find at most view sites and truck rests. Occasionally, descriptive information, toilets, and drinking water are provided at view sites.

190   Highways: An Architectural Approach

Ports of entry, rest stops, and welcome centers embrace a greater array of facilities and services. In designing a rest stop, the highway architect should consider not only the number of users, but also special needs and desires of users. Truckers, tourists, and elderly each require emphasis on different services. Truckers need adequate parking space together with no blockages that require them to back up their rigs. (Backing a turnpike double in a crowded space that is constantly being crossed by pedestrians can lead to more unpleasantness than just a few hundred well-chosen expletives on the part of the vehicle operator.) Tourists like to have map boards, brochures, and other handy information related to the immediate locale. Elderly travelers require more attention to facilities for disabled as well as fewer steps and shorter walking distances. Each rest stop and welcome center design ought to take the user mix as basic in determining extent and type of facilities and services, using agency and federal standards as guidance rather than dogma.

Since available money is the prime mover in determining the extent of facilities and services, the architect may wish to use the following generalized list of most-to-least desired items and place a cut off at a point if and where programmed funding comes out short when making a preliminary estimate of costs:

- Toilets (80 to 90 percent of rest stop users);
- Telephone (particularly emergency non-coin operated);
- Drinking water;
- Lighting;
- Maps and displays;
- Vending machines;
- Commercial information (food, lodging);
- RV dumpsite;
- Picnic tables;

**FIGURE 14-7.** Drinking water and toilet facilities are regarded by the public as two of the most important features of a rest stop.

- Pet exercise area; and
- Additional amenities suited to the locale.

## Upgrading Existing Sites

As previously mentioned, view sites are sometimes converted to rest stops when traffic increases to the extent an upgrading is justified. Also, if an existing rest stop shows signs of overuse, its facilities may need to be expanded or supplemented. However, if water supply, electricity, or sewage disposal is limited on-site, the better alternative is probably to develop an additional brand new rest stop some distance (30 miles or so) down the road.

## Special Concerns

Various surveys have demonstrated that visibility from the highway is an important asset for any rest area. Travelers want to know what to expect. A sign that simply says "rest area next exit" has come to be distrusted by many travelers when they cannot actually see the rest area as they approach.

Remote rest stops that do not get regularly patrolled by the constabulary can be a problem in some parts of the country. Florida, for example, has closed down and removed a number of rest stops because of illicit activities taking place at the stops.

The biggest single consideration continues to be maintenance of the facilities at a rest area. When trash cans are not emptied regularly, toilets are vandalized and not repaired, lights bulbs are broken and not replaced, graffiti is rampant, and litter is abundant, a rest area ceases to serve its purpose. It behooves the architect to design rest areas with ease of maintenance a principal concern.

**FIGURE 14-8.** Lighting, maps, and displays are also deemed important by travelers.

## INTERMODAL TRANSFER

There continues to be excessive inertia by planning people when it comes to developing intermodal facilities—not so much in the transfer of commodities, but considerable when it comes to moving people. This seeming paradox appears repeatedly in the literature. As an example, in a paper *Critical Factors in Planning Multimodal Passenger Terminals*, which appears in *TRR 1221*, the following anecdotal reference was cited:

> ...efforts to develop surface passenger terminals for bus and rail have been uncoordinated in Canada due to a lack of incentive to combine efforts. Each carrier prepares its own plans without consulting others. An incident from Saint John, New Brunswick, in 1979 provides a good example. VIA RAIL consolidated a former Canadian National railway station and a Canadian Pacific railway station, located in the suburbs, into one downtown location. While VIA was preparing its plans, SMT, the major regional intercity bus carrier, was preparing plans for its own terminal at another location only a short distance away. This example illustrates a missed opportunity.

A missed opportunity? That's an understatement, unless we're discussing a missed opportunity for a tar and feathering ceremony featuring at least two planners! And it should not be necessary to point out that this lack of coordination between planners working in different transportation modes is not limited to Canada.

A contrast to the above experience is shown in the accompanying schematic plan (Figure 14-9) of a consolidated transportation center in Westchester County, New York. Multimodal transit centers are becoming more in vogue such that coordination of transfers to and from automobiles, buses, and heavy and light rail is getting to be feasible. Most of the intermodal transfers, however, continue to be from one single mode to another single mode. *TRR 880* contains two case studies (pages 27-38) that may help to guide the architect in certain aspects of intermodal transfer. In *TRR 925* (pages 21 and 22) there are sketches of what an intermodal facility might embrace. Design elements for small installations can be gleaned from information in the U.S. Army Coastal Engineering Research Center *Report SR2*.

### Airport Terminals

Long, time-consuming, luggage-toting treks from parked cars to boarding gates are a bane for many travelers. Moving sidewalks at many major terminals help somewhat and automatic people mover (APM) systems, where installed and operating, help even more. Terminal configuration has much to do with the total walking distance required of boarding and deplaning passengers. Yet the highway architect may be called upon to design intermodal transfers at airport terminals that rely

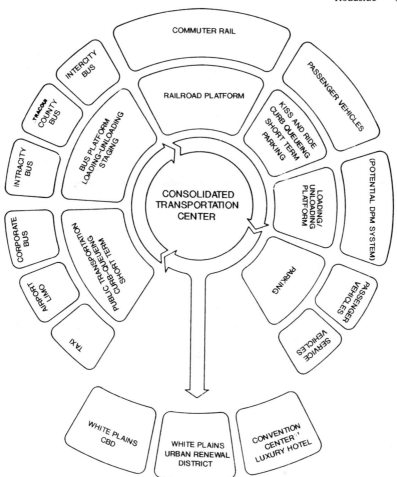

**FIGURE 14-9.** Schematic of a proposed intermodal terminal north of New York City. (From Transportation Research Record 877.)

solely on shuttle buses. Before beginning a design saddled to an outdated, inadequate concept, the architect ought to investigate various alternatives that enlist up-to-date technologies for getting people where they want to go.

Air cargo continues to climb in its share of commodity carrying. Most air terminals have separate areas dedicated to freight transfer, although more and more passenger flights are being required to increase their emphasis on cargo operations.

## Railroads

Although passenger access to Amtrak and other passenger trains seldom poses a problem, ancillary items such as baggage handling require some measures to enhance travel convenience. There are times when design of an intermodal auto-train facility can streamline or smooth out the transfer of people and their accessories during the short period that the train spends in the station.

Intermodal transfer of goods (truck-freight train) has become a much more sophisticated operation. Double-stack container cars coupled with continuing evolution of high-volume transfer handling equipment place greater emphasis on efficient access and clearing of trucks and semis at the transfer location. A highway architect who is called upon to help design a freight terminal might best seek expert consultation.

## Ports and Waterways

Ferry boats could be the first thought of many when they consider water transport. Access from shoreside to the ferry piers, however, is not always accomplished with a minimum of stress to the automobile occupants. Highway architects may well consider providing some simple amenities in situations where queues of vehicles must wait while the drivers and passengers patiently—well, not always patiently—pass the time in a rather drab setting. A little imagination can improve the setting so as to make the time spent waiting to board the boat more useful.

Movement of goods, particularly on our inland and intracoastal waterways, requires efficient transfer from truck (or rail) to barges and boats and ships. This topic is much too broad to develop in this chapter; the architect should be aware, though, that the function of the road network in the vicinity of intermodal transfer

**FIGURE 14-10.** Waiting for the ferryboat to arrive can be a dreary experience unless some interesting roadside amenities are provided.

ports ought to be worked out in consultation with those responsible for the operation of the intermodal facility.

Increases in land-bridge service, where goods are unloaded at an ocean port from a ship too large to negotiate the Panama Canal and trucked (or "railroaded") across country to another ship provides benefits to our internal transportation industry. Transfer of goods that require import and/or export licensing may mean that garages or warehouses are to be included in an intermodal master site plan in order to allow for the time that containerized goods are required to wait for consummation of the bureacratic process.

## COMMUTER PARKING LOTS

Maximum, or at least optimum, use of a commuter parking lot is somewhat dependent on its location. Nevertheless, a well thought out and properly designed park-and-ride facility can be very effective in making it well used. Design elements include adequate ingress and egress during peak periods, together with satisfactory capacity of the local road feeders. Additionally, comfortable and efficient entry onto and exit from the arterial leading to the high density employment center is highly desirable—let's make that mandatory rather than just desirable.

Where a remnant piece of land adjacent to an expressway interchange is designated (sometimes informally) as a commuter parking lot and, as often happens, the bottlenecks and delays in getting in and out of the lot add up, a proportionate number of one-time users become discouraged and the commuter parking lot receives less than optimum/expected usage. A series of well-designed and therefore well-used commuter lots can be one means toward reducing the need for an excessive investment to enlarge arterial highways, expressways and freeways leading to and from high density employment centers.

What are the important design elements that need to be considered?

Let's first of all pursue this from the vantage point of the user. Separate entry and exit is preferred for diverting (as opposed to concentrating) flow. Autos attempting to get into the lot at the same time and place as others attempting to get out can be reminiscent of the swinging door in a barroom scene where Wyatt Earp is offering free drinks for his friends and lead slugs for his enemies. Such a condition points dramatically toward the provision of separate entry and exit. Psychologically too, few impatient commuters like to turn around and backtrack to the place where they entered the lot; it undoubtedly stifles their inherent striving to make progress. If a site is to be a rendezvous, the pool driver wants to keep going. In order to accomplish this effectively, the pickup points must be selected for comfort and convenience. This means easy accessibility, protection from inclement weather for those waiting for their pool car (or van or bus), and having pickup and discharge points positioned so that when poolees become temporary pedestrians they are not forced to dart across the paths of other commuters' autos (which often give the impression of making practice runs for a TV chase sequence).

The accompanying sketch shows schematically one approach to accomplish our design goals (see Fig. 14-11). But look out, here come the site planners!

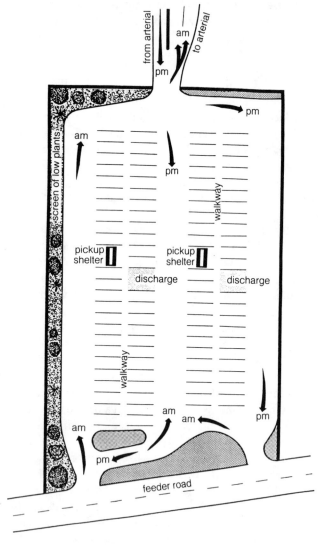

**FIGURE 14-11.** A suggested commuter parking lot layout.

"Hey, that's a helluvalot of wasted land. We can get oodles more parking stalls into that space! I'll bet we could save $20,000 by a more effective use of the land."

Undoubtedly true, but our objective here is not to provide the most number of parking spaces, but rather to provide a high-use facility that can help to reduce the need to add another lane or two on an existing arterial leading to a high-density employment area. If it's necessary to spend an extra $20,000 so as to save $100,000, the cost-effectiveness is obvious. The goal is to make the commuter parking lot a well-used item. A few low-cost amenities may well do the job that a bare-bones approach would not. Some amenities that could be considered include:

Public telephone(s)
Maildrop
Vending machines for newspapers, food, and drink
Trash receptacles
Information system regarding weather and bus schedules
Attractive and understandable signing
Sheltered bicycle racks
Planting or screening

Local behavior patterns and driving idiosyncracies weigh heavily in developing the layout. Each situation needs to be judged from the criterion that potential usage of the commuter parking lot outweighs other considerations imposed when applying traditional standards for parking lots.

## OTHER ROADSIDE MATTERS

Guardrails, delineators, and signs are cited by traffic engineers as the most important "roadside" items. They are important, but belong more rightly under the category of operational fixtures, as part of the roadway itself. Chapter 21 will deal with guardrails, delineators, and signs.

Some roadside issues have already been addressed in preceding chapters. Depending on the nature and scope of a project, roadside design could constitute a major portion, or be miniscule, or, more than likely, something between the two extremes.

**References**
AASHTO. *A Guide for Transportation Landscape and Environmental Design*. 1991.
    Washington, D.C.: American Association of State Highway and Transportation Officials.
AASHTO. *A Policy on the Accommodation of Utilities Within Freeway Right-Of-Way*. 1989.
    Washington, D.C.: American Association of State Highway and Transportation Officials.
AASHTO. *A Strategy for Research, Development, and Technology Transfer—Intermodal*

*Issues*. 1989. Washington, D.C.: American Association of State Highway and Transportation Officials.

Adler, Bernard, and Marvin C. Gersten. 1982. Planning and preliminary design of White Plains, New York, transportation center. *Transportation Research Record 877*:114-122.

Bell, David W. R., and John P. Braaksma. 1989. Critical factors in planning multimodal passenger terminals. *Transportation Research Record 1221*:38.

Bendtsen, Hans. 1991. Visual principles for the design of noise barriers. Paper read at 70th Annual Meeting of the Transportation Research Board, January 13-17, 1991, Washington, D.C.

Bowler, Charles E., et al. 1986. *Park-and-Ride Facilities—Guidelines for Planning, Design and Operation*. McLean, Virginia: Federal Highway Administration.

Bullard, Diane, and Dennis L. Christiansen. 1981. *Park-and-Pool Facilities Survey Results and Planning Data*. College Station: Texas Transportation Institute.

Bullard, Diane, and Dennis L. Christiansen. 1983. *Guidelines for Planning, Designing and Operating Park-and-Ride Lots in Texas*. College Station: Texas Transportation Institute.

Coastal Engineering Research Center. 1974. *Small-Craft Harbors: Design, Construction, and Operation, Report SR2*. Washington, D.C.: U.S. Army Corps of Engineers.

FHWA. *Evaluation of Deer Mirrors for Reducing Deer-Vehicle Collisions, Report RD-82-061*. 1982. Washington, D.C.: Federal Highway Administration.

FHWA. *Wildlife Considerations in Planning and Managing Highway Corridors, Report TS-82-212*. 1982. Washington, D.C.: Federal Highway Administration.

Fowler, David W., et al. 1987. *Design Recommendations for Rest Areas*. Austin, Texas: Center for Transportation Research.

*Freight Facts*. 1989. New York: The Port Authority of New York and New Jersey.

Galligan, Donald C., Jr., 1990. Land Use Controls and Policy for Airport Development. *Transportation Research Record 1257*:1-9.

Herman, Lloyd. 1991. Full scale testing of single and parallel highway noise barriers. Paper read at 70th Annual Meeting of the Transportation Research Board, January 13-17, 1991, Washington, D.C.

Hunsucker, D. Q., G. W. Sharpe, and R. C. Deen. 1987. *Design and Performance of Highway Shoulders*. Lexington: Kentucky Transportation Research Program.

Jessup, D. R., G. Van Wormer, and H. Preston. 1983. *Guidelines for the Design of Transit Related Roadway Improvements*. St. Paul, Minnesota: Metropolitan Transit Commission.

King, G. F. 1989. *NCHRP Report 324, Evaluation of Safety Roadside Rest Areas*. Washington, D.C.: Transportation Research Board, National Research Council.

Klatt, Richard T., Mary S. Smith, and Majid M. Hamouda. 1987. Access and circulation guidelines for parking facilities. In *Compendium of Technical Papers*, pp. 310-314. Washington, D.C.: Institute of Transportation Engineers.

Leccese, Michael. 1989. Roadways Recovered. *Landscape Architecture* 79(3):49-55.

Melton, William, Anh Tran, and Julie Leverson. 1989. *Rest Area Usage Design Criteria Update*. Olympia: Washington State Department of Transportation.

NCHRP Digest 161. 1987. *Public and Private Partnerships for Financing Highway Improvements*. Washington, D.C.: Transportation Research Board, National Research Council.

NCHRP Digest 309. 1988. *Protection of Pipelines Through Highway Roadbeds*. Washington, D.C.: Transportation Research Board, National Research Council.

NCHRP Synthesis 20. 1973. *Rest Areas*. Washington, D.C.: Transportation Research Board, National Research Council.
NCHRP Synthesis 115. 1984. *Reducing Construction Conflicts Between Highways and Utilities*. Washington, D.C.: Transportation Research Board, National Research Council.
Noel, Errol C. 1989. Implementation and operation of park-and-ride lots. Urban traffic systems and operations. *Transportation Research Record 1232*:76-83.
Pacelli, A. J. 1980. Intermodal Transportation Garage Facility. *ASCE Journal of Transportation Engineering* 106(4), 401-413.
Perry, Kirby W., *et al.* 1987. *Design Manual for Rest Area Comfort Stations*. Austin, Texas: Center for Transportation Research.
Schafer, James A., Stephen Penland, and William P. Carr. 1985. *Effectiveness of Wildlife Warning Reflectors in Reducing Deer-Vehicle Accidents in Washington State*. Transportation Research Record 1010:85-88.
Schneider, Jerry B., *et al.* 1980. *Planning and Designing a Transit Center Based Transit System: Guidelines and Examples from Case Studies in Twenty-two Cities*. Washington, D.C.: Urban Mass Transportation Administration.
State of the Art Report 4. 1986. *Facing the Challenge, the Intermodal Terminal of the Future*. Washington, D.C.: Transportation Research Board, National Research Council.
Transportation Research Record 880. 1982. *Port Development, Waterway System Analysis, and Marine Transit*. Transportation Research Board, National Research Council.
Transportation Research Record 913. 1983. *Roadside Vegetation Restoration and Protection*. Washington, D.C.: Transportation Research Board, National Research Council.
Transportation Research Record 925. 1983. *Inland Water Transportation*. Washington, D.C.: Transportation Research Board, National Research Council.
Transportation Research Record 970. 1984. *Safety Appurtenances and Utility Accommodation*. Washington, D.C.: Transportation Research Board, National Research Council.
Transportation Research Record 1075. 1986. *Roadside Design and Management*. Washington, D.C.: Transportation Research Board, National Research Council.
Transportation Research Record 1160. 1988. *Traffic Control Devices*. Washington, D.C.: Transportation Research Board, National Research Council.
Transportation Research Record 1222. 1989. *Ports, Waterways, and Inland Water Transportation*. Washington, D.C.: Transportation Research Board, National Research Council.
Transportation Research Record 1263. 1990. *Ports, Waterways, Rail, and International Trade Issues*. Washington, D.C.: Transportation Research Board, National Research Council.
Transportation Research Record 1273. 1990. *Airport Terminal and Landside Design and Operation*. Washington, D.C.: Transportation Research Board, National Research Council.
Transportation Research Record 1279. 1990. *Hydrology and Environmental Design*. Washington, D.C.: Transportation Research Board, National Research Council.
Wirasinghe, S. C., and U. Vandebona. 1987. Passenger walking distance distribution in single- and dual-concourse centralized airport terminals. *Transportation Research Record 1147*:40-45.
Zeigler, A. J., *et al.* 1986. *Guide to Management of Roadside Trees, FHWA-IP-86-17*. McLean, Virginia: Federal Highway Administration.

# 15
## Right-of-Way

In the design of a new highway—or in the redesign of an existing highway segment—there may be a need to obtain additional real estate. Also, in an increasing number of cases, there is a likelihood that some property may be transferred from transportation to some other land use. In the former case, the procedures for obtaining highway rights-of-way have been established and ongoing for many years. In the latter case, disposing of lands, the issue is still in a form of transmogrification. In some states and larger municipalities, it is "illegal" for a highway agency to dispose of land holdings directly: "The highway department is not in the real estate business." Other state and local governments have policies that permit land exchanges and selling of no-longer-needed highway properties without an abundance of paperwork. An additional item to contend with is the FHWA position with regard to properties originally purchased using federal funds.

Something else to be cognizant of when federal money is being used for any portion of a project is the *Uniform Relocation Assistance and Real Property Acquisition Policies Act of 1970*. In Title III of the Act, the U.S. Congress has stipulated many structured requirements regarding compensation to and relocation of property owners when additional land is needed for transportation purposes.

### ACCESS

Perhaps the most often encountered quagmire is the desirability of converting *unlimited access* to *limited access*, *controlled access*, or *non-access* for enhancement of highway capacity, efficiency, and safety. Political and local economic (business) pressures often forestall or obstruct any effort to get the optimal operation from a new or remodeled highway segment. A useful primer on access conversion is found in the NCHRP pamphlet *RRD 164*. The digest contains a bibliography and references for the architect who wishes (or is required) to go into the subject in greater depth.

## ROLE OF THE ARCHITECT

Although the highway architect (or designer) usually takes the initiative in determining right-of-way requirements, the actual appraisals, property surveys, condemnations (if necessary), and purchases are handled by a separate branch, usually of the sponsoring highway agency. Yet the architect should be somewhat aware of the appraisal procedure. In the case of business properties, income generated by the property usually receives the most attention. This means that some old, run down property might command a high value if the owner can convince the appraiser (without getting the IRS into the loop) that the property produces considerable revenue. Residential units, on the other hand, are looked at in light of their market value: dilapidated dwellings don't demand dollars. Some appraisers prefer to itemize and total all the costs involved (land, replacement, depreciation, etc.) and then wrangle with the property owner.

Severance damages, where the remainder property has little or no value to the owner, can be a sleeper in determining initially how much to allocate for property acquisition. Many times, severance damage costs (including legal and time delay) deem it advisable to have the highway agency buy the whole parcel and then decide what to do with the unneeded real estate (e.g., multiple use/joint development). Also, if some of the acquired land is found to contain hazardous waste, the highway agency is responsible for proper disposal—and this can be expensive. What it all means is that the architect should be careful about making any quick guesses on how much the various property acquisitions are going to increase the cost of the project.

Since responsibilities of the highway architect in right-of-way acquisition vary greatly from one section of the country to another, a basic concern has to be the judgments used for making prudent decisions. Most of these judgments have already been made during both planning and preliminary design phases.

Quite often, late in the design phase, the architect may be called upon to prepare a tract map showing properties and portions of properties to be acquired for highway purposes. Property maps are superimposed upon final plan sheets with the extent of each taking clearly marked and dimensioned. Descriptions and deeds for each parcel are usually prepared to accompany the tract map.

Deeds? Well there are three main types of deeds: Quitclaim, Warranty, and Easement. A *quitclaim deed* transfers an owner's interest in a particular piece of property; no guaranty is made by the owner of a good and merchantable title. A *warranty deed* is a document where the grantor (owner) warrants the title against defects (other people's claims on the property) that may exist or may arise at any future time. An *easement* only allows the use of the property (usually land) for a specific purpose without transferring actual ownership of the land.

**Fulfillment**

Completion of the task to be performed here by the highway architect is entirely dependent on the requisites of the sponsoring agency. It could include complete maps, surveys, descriptions, and deeds of all parcels involved, or a simple right-of-way taking line on the set of final plans (whereupon another division of the sponsoring agency will perform the detail work, usually in their own time-frame).

**References**

Levin, David R. 1982. Right-of-way acquisition. In *Handbook of Highway Engineering*, ed. Robert F. Baker, L. G. Byrd, and D. Grant Mickle, pp. 329–339. Malabar, Florida: Robert E. Krieger Publishing Company.

NCHRP Legal Research Digest 18. 1991. *Supplement to Planning and Precondemnation Activities as Constituting a Taking Under Inverse Law*. Washington, D.C.: Transportation Research Board, National Research Council.

NCHRP Research Digest 19. 1991. *Continuing Project on Legal Problems Arising Out of Highway Programs*. Washington, D.C.: Transportation Research Board, National Research Council.

NCHRP Research Results Digest 164. 1987. *Rights of Abutting Property Owner Upon Conversion of Uncontrolled-Access Road Into Limited-Access Highway*. Washington, D.C.: Transportation Research Board, National Research Council.

# 16
## Plans, Specifications, and Estimates

There comes a time to bring it all together. Probably, some elements of the design (e.g., surveys) were done some time ago. Therefore, updating of all the information collected and incorporated into the various aspects of the nearly complete design is mandatory. Except for certain features to be covered in Chapter 21, the preceding chapters have outlined the basics for a complete, typical, fairly complex highway project. Coordinating all these facets into a single design package is the responsibility of the highway architect (regardless of the architect's title or job description).

Highway projects contain a distinct adversarial nature. Back in Chapter 12, the *Design and Build Procedure* was alluded to. Many building and site development endeavors are put together this way; the same outfit that does the planning and design gets to build (and sometimes operate) the edifice or project. Highway work, on the other hand, is done by contractors who, commonly, specialize in construction only. They get a set of plans and specifications and then do their thing. When they are finished, they turn the completed facility back to the owner/agency. It has worked well for more than a century, but the system does have a bundle of pitfalls.

A highway agency will almost always take the low bid from those submitted by contractors. Consequently, the agency has to monitor (watch over) all activities performed by the low-bid contractor. In turn, the contractor, desiring to make a profit on the work, tries to keep costs down by providing the minimum acceptable work. Although cooperation between the contractor and the agency representatives (field engineers and inspectors) is generally the rule, there are many instances where adversarial confrontations arise. They arise from poor sets of plans, varying interpretations of specifications, and inaccurate estimates of quantities.

## PLANS

Acknowledging local practice is the key to turning out a successful set of plans. Referring to recent drawings for similar-type projects by the sponsoring agency and/or by other nearby highway agencies, the architect can be afforded a glimpse at acceptable techniques and formats, including:

- Title page;
- Sheet layout;
- Index maps;
- Page sequence (i.e., plan views, drainage, structures, etc.);
- Extent of detailed drawings;
- Summary sheet arrangement;
- Use of standard drawings;
- Applicable schedules;
- Descriptions of traffic phasing and control;
- Scales and legends normally used;
- Structure sheet organization; and
- Traffic signal specialties.

Any features on the reference plans that appear weak, such as poor drafting style, inconsistencies from one sheet to the next, faulty or too small lettering, indistinguishable markings, and the like, are features that the architect need not imitate. Final plans should display an attention to detail and an image that states that a great deal of care went into their preparation. Investment in a high-quality plotter is now common to most design offices.

Occasionally, "local practice" gets in a rut. The architect may find it prudent to solicit help from experienced field personnel who have supervised project construction from either the highway agency's or the contractor's perspective, or both. Having such persons review the finished plans can often turn up many useful suggestions. Some of the particulars that continually vex contractors, field engineers, and inspectors include:

- Work items described on the plans that have no corresponding items in the proposal;
- No thought given as to how the project could be expeditiously constructed;
- Contradictory information from one sheet to the next (e.g., plan sheets show one thing, drainage sheets show something different); and
- Inadequate materials investigation and identification.

And yes, there are others.

## Safety Review

On most large projects involving federal aid highway funds, a safety report is required. Whether required or not, a review of the plans to check safety angles is a good idea. For openers, the architect may wish to define or isolate those safety features that are of greatest concern. Some major problems and identified hazards can be found in the list on page 9 and photos on pages 15-33 of *NCHRP Synthesis 128*. Seven rating levels of rural roadside hazards are shown in photos on pages 8-14 of *FHWA/RD-87/094*. Further investigation could lead to the seventeen examples of hazardous elements listed on page 7 of *NCHRP Synthesis 132*. Then, for added emphasis, the architect may find it enlightening to flip through the pages of *TRR 1133*; he or she will come across quite a few graphic photos of just what happens when errant vehicles impact various pieces of roadside furniture and obstacles.

Although geometric guidelines and most design standards contain safety as a built-in, there are many situations that can arise to cause safety concerns once a project is opened to traffic. For this reason, the architect should review design decisions, particularly where larger trade-offs are involved, to assure that overall safety has not been compromised. There are also a few design initiatives that offer certain safety benefits:

- Headlight beams from approaching traffic on divided highways can be minimized by median plantings or by glare screens described on pages 3-7 of *NCHRP Synthesis 66*. Requirements of screening are discussed on pages 12-16 of the same document.
- As freeways, expressways, and other non-access highways become more crowded, a major safety concern is the backing up of traffic onto the main line at exit terminals. Some particulars relating to this concern are explained on pages 22-28 of *NCHRP Synthesis 35*.
- Designing road improvements, in the sun belt particularly, must heed the reality that the average age of drivers is going up. *TRB Special Report 218*, volume 1, pages 4-6, indicates four domains where existing roadway design standards may not be ideal if a significant percentage of drivers and pedestrians are senior citizens.
- Trees are often troublesome. Accidents involving roadside trees in Michigan are the subject of a paper presented on pages 37-43 of *TRR 1127*.
- Evident to most automobile owners, the sizes of vehicles keep changing. To anticipate the future, it's worth considering what may be in store in the realm of vehicle characteristics. Autos will probably get smaller while trucks will probably get bigger (see Fig. 16-1). Some educated crystal ball studies can be found on pages 108 and 109 of *TRB State of the Art Report No. 6*.
- Finally, in the battle for the bucks, a high priority must go to Value-Architecture.

206   Highways: An Architectural Approach

Achieving "roadway consistency" on a highway segment is probably the optimum means of employing the principles of Value-Architecture as they relate to safety. Sixteen type-improvements that have proved successful in Iowa are highlighted in *NCHRP Synthesis 132*, pages 10-13.

Nevertheless, a set of final plans should be subjected to the safety review. On major projects, such a review is commonly undertaken by a staff of experienced safety professionals. However, if the architect is given personal responsibility for conducting a final safety review, there are several key items that require attention on almost every project:

- Consistency of layout within intersections;
- Compliance with applicable provisions of the *AASHTO Green Book, Manual on Uniform Traffic Control Devices (MUTCD)*, and *AASHTO Guide for Selecting, Locating and Designing Traffic Barriers*;
- Appropriateness in level of striping and traffic control devices (an extension of compliance with *MUTCD*);
- Assurance that existing dangerous conditions at high-hazard locations have been removed or lessened; and
- Determination that no new hazard sites have been introduced.

An elementary means to consider these latter three facets is for the architect to "role play": to take an imaginary trip in both directions through the completed project (with the aid of the nearly final plan sheets) and determine where any hazardous locations might be; then to do the same as a handicapped pedestrian attempting to get across the highway; also a bicyclist negotiating turns in traffic; and so forth. Some clues to help in these latter two roles can be found on pages 116-119 of the *Handbook of Highway Safety Design and Operating Practices*.

**FIGURE 16-1.** Although autos are getting smaller, predictions are that trucks will get bigger.

## Quality of Plans

Reviewing, checking, and comparing to specifications rarely get overdone. Ethical bidders do desire to remain competitive. They like to "c" a good set of plans: clean, complete, consistent, clear, compliant (with applicable regulations), coherent, credible, correct, coordinated, and consummate. A good set of plans does a lot toward obtaining lower bid prices, reducing legal liabilities, completing the design concept, and having the project built in a timely manner.

## SPECIFICATIONS

Plans provide the pictures for the work to be done on a project. Specifications provide the narrative or captions that go with the pictures. Highway engineers, with a traditional disdain for the written word, usually wait until the last minute before beginning the assembly, editing, and writing of specifications for the various contract items. ("Maybe I can get someone else to do it.") Architects, be forewarned.

Most of the heavy construction projects nationwide (dams, buildings, site developments, and the like) are governed by specifications that follow the Construction Specifications Institute (CSI) format. Not so the highway community. Although a few local government highway agencies have opted for the CSI format, almost all state highway agencies (at this writing) observe their own favored specification format. This situation is apt to transmogrify soon.

Highway construction contracts are usually bid on a per-item basis. Each item, therefore, requires a corresponding specification. A specification will almost always describe the material(s) to be used, the method of incorporating the material, and how the completed work will be measured and paid for.

Simple?

Sometimes—at least in cases where there is a standard agency specification, competent inspection, and an ethical contractor. Alter or modify any one of these essential ingredients and the likelihood of trouble goes up several notches. Alter or modify two and...(groan). The designer, having control only over the specifications, can hope that, by fussing over the assembly, editing, and writing of the various specifications, at least one of the essential ingredients will not cause trouble.

However, if the architect's duties carry over into the construction phase (and they should), observation of and subordination to the intent of each specification can be encouraged if not assured.

### Standard Specifications

Hopefully, the sponsoring agency has a book of standard highway construction specifications, or is willing to use another nearby agency's book as a surrogate. One should apply it to the maximum extent feasible. If in a bind, the AASHTO *Guide Specifications for Highway Construction* can be used as the base document.

**Special Provisions**

Supplementing the references to the items contained in the standard specifications, most projects are also attended by a number of unique specs or special provisions. In cases where a specialty item (or method of construction) is required and is not covered in the standard specs, the next best thing is to locate another highway agency spec book and see if it (the specialty item) is contained therein. If not, it may be necessary to go to a supplier, distributor, or manufacturer who handles the item in question. Most will gladly provide a specification—so worded that it probably excludes all other competitive products. Consequently, a second opinion ought to be sought. What the architect may wind up with is two contradictory, mutually exclusive manufacturer-supplied specifications describing essentially the same product.

What to do?

Applying the same format and style as used in the standard spec book, use the information contained in the two (or more) manufacturer specs and piece together a draft spec. Run it back through the manufacturers and/or suppliers for their comments. Eventually, maybe after several iterations, an architect can develop an acceptable directive to assure that:

- The contractor understands what is wanted;
- Responsibilities of the field inspector are indicated;
- Standards implicit in the design are met;
- Legal issues are properly addressed;
- Method of measurement is adequately described; and
- Manner of payment to the contractor is made clear.

There may be a lesson here. If so, it would strongly suggest that (1) the architect ought to begin assembly of the specifications at an earlier stage than the end of the job, and (2) standard specifications should be used wherever feasible.

Some specifications writers have been successful in pirating or adapting specialized specifications from previously constructed projects. (It can be legal when the spec is not copyrighted.) Such practice can be problematical, though (particularly when describing geo-textiles), because the function that a product served in a previous project might not be the same as that desired for the proposed project.

To supplement standard specifications and special provisions, the architect could decide to use some nationally recognized institutional standards and specifications. The more commonly used ones are available from the following organizations:

AASHTO
American Association of Nurserymen (AAN)

American Concrete Institute (ACI)
American Institute of Steel Construction (AISC)
American National Standards Institute (ANSI)
American Society of Mechanical Engineers (ASME)
American Society for Testing and Materials (ASTM)
American Wood Preservers Association (AWPA)
American Water Works Association (AWWA)
Concrete Reinforcing Steel Institute (CRSI)
Institute of Transportation Engineers (ITE)
National Association of County Engineers (NACE)

The beauty of using standards or specifications from the above organizations is that it is only necessary to cite their spec number (e.g., ASTM *B 241* or AASHTO *M 154*), rather than including the entire text. The problem with using the above is that the architect has to sort through an immense amount of paper to find what he or she may be looking for—and then maybe not find it.

## Value Engineering

To provide an incentive to a contractor to initiate and develop cost savings proposals, a number of highway agencies now permit modifying the specifications by the contractor during the course of work. Although a supplemental agreement is usually executed for any Value Engineering proposals submitted by the contractor after construction of the project has begun, the architect may be called upon to make judgments as to the applicability of the contractor's proposal.

## Outside Review

If the architect has decided to solicit aid in reviewing the final plans, he or she might also wish to extend the solicitation to reviewing the specs. Field personnel may point out where incomplete specifications could lead to overruns and extra payments, where available construction materials would not meet specs, or where time constraints could limit the number of contractors willing and able to bid on a project.

Frequently, an overly cautious highway agency tries to protect itself by inserting a penalty, or liquidated damage, clause in the contract documents. Rarely do such clauses do any more than raise the bid prices.

# ESTIMATES

Average bid prices are readily available (most often on a data base at the state highway agency) for the previous year on all standard items. An inflation factor can be added on by adapting the Consumer Price Index (CPI) or Producer Price Index

(PPI) kept current by the U.S. Department of Labor. Factors for project size and location within a jurisdiction are also generally applied.

Some—only some—of the more common items a designer is likely to run up against are the following:

| | Unit of Measure |
|---|---|
| **Site preparation items** | |
| Clearing and grubbing | Lump sum or acre |
| Removal of buildings and structures | Each |
| Removal of curbs, gutters, and sidewalks | Linear feet (LF) |
| **Earthwork items** | |
| Stockpiling of topsoil | Cubic yard (CY) |
| Excavation (which usually includes placing it in fills) | CY |
| Imported borrow | CY |
| Granular borrow | CY |
| Preparing pavement subgrade | Square yard (SY) |
| Excavation for bridges | CY |
| **Pavement items** | |
| Subbase | CY |
| Base course | Ton or CY |
| Surface course (asphalt concrete—not including bitumens) | Ton or SY |
| Bituminous material (additive to surface course) | Gallon or ton |
| Portland cement concrete—not including cement | CY or SY |
| Portland cement (for the concrete) | Barrel (Bbl) |
| Reinforcing steel | SY |
| Joint sawing | Linear feet (LF) |
| Curb and gutter | LF |
| **Drainage items** | |
| Trench excavation | LF |
| Reinforced concrete pipe (RCP)—of many sizes | LF |
| Corrugated metal pipe (CMP)—also many sizes | LF |
| Headwalls | CY or each |
| End sections | Each |
| Drop inlets, manholes, and catch basins | Each |
| Riprap | CY or SY |
| **Structure Items** | |
| (These would normally be specified by the designer of the structures.) | |
| **Roadside items** | |
| Sidewalks and driveway aprons | LF or SY |
| Fencing | LF |
| Spreading topsoil | CY, SY, or Ton |
| Seeding | Acre or pound |
| Landscape plantings | Each |
| Utility and railroad items—specified by the respective companies | |

Traffic operation and safety items
- Guardrail (guide rail) — LF
- Median barriers — LF
- Delineators — Each
- Signs — SF or each
- Pavement markings and striping — Each or LF
- Maintenance and protection of traffic — Lump sum
- Lighting and signal items—usually specified by specialists

Nonstandard items can be estimated in a number of ways, the most common practice being through suppliers and manufacturers. The same caveats and modi operandi are suggested, as previously outlined above under "Special Provisions."

Contractors are ever on the alert for obvious errors in quantities. They counter these errors by unbalancing their bids: putting in a high unit price for any item that has been estimated too low; putting in a low unit price for an item that has been grossly overestimated.

Compliance both with federal guidelines contained in the appendix to *FHPM 6-3-3-1* and with recommended procedures outlined in *T-5080.4* is mandated for federal aid projects.

## FINAL PACKAGE

When it's all complete and checked, the architect (or supervisor) wraps up the design package (plans, specifications, estimates, calculations, etc.) and transmits it for the sponsoring agency's review.

## POSTMORTEMS AND REFLECTIONS

A prime reason for an individual to go into the field of architecture is to do something innovative and creative, to try new things and methods. After going through the design phases of a highway project, a highway architect may well feel jaded by the constraints and restraints placed upon trying anything different, discouraged by attempting any unique approach to the standard way of doing things.

### Innovation

Creativity and imagination have long had a minuscule niche in highway technology. (After all, the Appian Way held up pretty well; why change the way we do things?) Highway architects will find barriers to adoption and use of new concepts, products, and methods extremely formidable. Crocodile-infested moats, tank traps, land mines, legions of entrenched bureaucrats, and irreversible regulations line the treacherous path that one must take in order to gain acceptance of a worthwhile

212    Highways: An Architectural Approach

**FIGURE 16-2.** Innovations in highway design takes a long time to get into production. The single point urban interchange (described in Chapter 13) has taken 20 years to gain acceptance.

concept, product, or method that has not made it to the "approved list." This attitude has not gone unnoticed outside the highway community. According to a 1988 U.S. General Accounting Office (GAO) report *How State Agencies Adopt New Pavement Technologies:*

> Perspectives of decision makers within an agency might also influence acceptance of innovative technologies that require a change from past practices. Also, the likely reaction of employee or labor groups may affect whether organization officials try a new method or product. Any change that would result in loss of job rights or security may be avoided. In addition, local decision makers might avoid trying new technologies in pavement sections because they perceive an element of risk that the technology could fail and that this failure could result in public criticism or legal liability. Another source of public criticism may stem from the inconvenience and delays motorists experience during highway construction...Another barrier to technology adoption may be low-bid procurement, where the least expensive method or product could be selected without regard to improved performance or lower long term (life-cycle) costs.

Remove all the "mights" and "mays" from the above quote and the picture is painted quite accurately. Anyone wishing to pursue these issues may begin by reviewing the entire 121-page GAO Report to the House Committee on Public Works and Transportation.

Responding to this almost dreary perspective: by incorporating architecture into the highway process, there is hope for the future. As more and more persons with the architect's point of view get into the highway field, the barricade of rigid, traditional, and immovable standards may well crumble. When this happens, there can be many benefits to highway users and others who directly and indirectly pay for highways. As stated in the summary in *NCHRP Synthesis 149:*

The challenge to highway agencies is to find new ways to define design and construction objectives that provide quality control, competition, and opportunities for innovation.

## Legal Sandtraps

Once upon a time, highway designers were fairly immune from liabilities; they were exempt from *injuries resulting from negligent design*. Not any more. Too many court decisions in recent times have held that there is a responsibility on the part of the highway agency (and its designers) to come forth with a nearly impossible *perfect* design. Perhaps every person in every field of endeavor will soon find it necessary to purchase malpractice insurance!

## Clean Up

Now that the project design phase is over, the architect gets to clean up the debris left behind. By all means, save the design book or copy thereof! Also, carefully catalog (and preserve in a safe place) any specialized plan sheets used to develop the design concept, floppy discs, and other irreplaceable items. The frightening spectre of being called upon to testify in a court case related to the project—or the less frightening prospect of being involved in field changes during construction—must be contemplated. A highway architect who has all the relevant design information readily at-hand is a serene, unruffled highway architect.

## References

AASHTO. *A Policy on Geometric Design of Highways and Streets.* 1990. Washington, D.C.: American Association of State Highway and Transportation Officials.

AASHTO. *Guide for Selecting, Locating and Designing Traffic Barriers.* 1977. Washington, D.C.: American Association of State Highway and Transportation Officials.

AASHTO. *Guide Specifications for Highway Construction.* 1988. Washington, D.C.: American Association of State Highway and Transportation Officials.

Byrd, L. G. 1989. *NCHRP Synthesis 149, Partnerships for Innovation: Private-Sector Contributions to Innovations in the Highway Industry.* Washington, D.C.: Transportation Research Board, National Research Council.

FHPM 6-3-3-1, *Plans, Specifications, and Estimates.* Regularly updated. Washington, D.C.: Federal Highway Administration.

FHWA. *Handbook of Highway Safety Design and Operating Practices.* 1978. Washington, D.C.: Federal Highway Administration.

FHWA. *Manual on Uniform Traffic Control Devices.* Regularly updated. Washington, D.C.: Federal Highway Administration.

FHWA. *Safety Cost-Effectiveness of Incremental Changes in Cross Sectional Design—Informational Guide, RD-87/094.* 1987. Washington, D.C.: Federal Highway Administration.

Goldbloom, Joseph. 1989. *Engineering Construction Specifications.* New York: Van Nostrand Reinhold.
Jameson, Gregory W., and Michael A. Versen. 1989. *Site Details.* New York: Van Nostrand Reinhold.
NCHRP Legal Research Digest 6. 1989. *Impact of the Discretionary Function Exception on Tort Liability of State Highway Departments.* Washington, D.C.: Transportation Research Board, National Research Council.
NCHRP Synthesis 35. 1976. *Design and Control of Freeway Off-ramp Terminals.* Washington, D.C.: Transportation Research Board, National Research Council.
NCHRP Synthesis 66. 1979. *Glare Screen Guidelines.* Washington, D.C.: Transportation Research Board, National Research Council.
NCHRP Synthesis 128. 1986. *Methods for Identifying Hazardous Highway Elements.* Washington, D.C: Transportation Research Board, National Research Council.
NCHRP Synthesis 132. 1987. *System-wide Safety Improvements: An Approach to Safety Consistency.* Washington, D.C.: Transportation Research Board, National Research Council.
Technical Advisory T5080.4. *Preparing Engineer's Estimates and Reviewing Bids.* Washington, D.C.: Federal Highway Administration.
Transportation Research Record 1127. 1987. *Innovation, Winter Maintenance, and Roadside Management.* Washington, D.C.: Transportation Research Board, National Research Council.
Transportation Research Record 1133. 1987. *Roadside Safety Features.* Washington, D.C.: Transportation Research Board, National Research Council.
TRB Special Report 218. 1988. *Transportation In An Aging Society.* Washington, D.C.: Transportation Research Board, National Research Council.
TRB State of the Art Report 6. 1987. *Relationships Between Safety and Key Highway Features.* Washington, D.C.: Transportation Research Board, National Research Council.
Turner, O. D., and Robert T. Reark. 1981. *NCHRP Synthesis 78. Value Engineering in Preconstruction and Construction.* Washington, D.C.: Transportation Research Board, National Research Council.
U.S. General Accounting Office. 1988. *How State Agencies Adopt New Pavement Technologies.* Report to the Chairman, Subcommittee on Investigations and Oversight, Committee on Public Works and Transportation, House of Representatives.

# Part 4
# Construction

# 17
# Contracts and Risk Management

Certain aspects of road and highway construction are unique to the species. Typically: contractors bid on a project; the (responsible) low bid is accepted by the highway agency; the contract between agency and contractor is drawn up and signed; the contractor sets up shop on site (mobilizes); the highway agency assigns a field engineer along with a staff of inspectors and record keepers (virtually none of whom had any contact with the design phase); and then the project gets underway. The previously discussed plans, specifications, and estimates are the only bridge between the design concept and the final product. These plans and specifications, together with the bid prices that have replaced the initial cost estimates, are part of the total contract. Other absolutely essential parts of the total contract are the "boiler plate" platitudes and "whereases" included to appease the barristers.

For years standardized contracts have provided adequate vehicles to accomplish the goal of expanding and improving the nation's highway network. However, beginning in the 1980s, a subtle movement towards slowly splitting the seams of these contracts, mainly on the part of contractors, has led to much hand wringing and embranglements within highway agencies. In some parts of the country, contract cost overruns are becoming epidemic.

The reasons behind the turmoil?

Some sources cite the change in emphasis from new construction to rehabilitation of existing highways, which leads to greater project complexity; thus, wider variations in interpretation of wording and intent of specifications becomes inevitable. Another theory is that, with advancing technology, highway agencies are not able (or willing) to accept better materials and more efficient methods proffered by contractors. A third opinion states that the old guard is on the way out. Those engineers, designers, and inspectors who got on board the highway express in the

1950s, when the big push to build the Interstate System was inaugurated, are now at retirement age. The young Turks replacing their ranks haven't yet learned to cope with crafty, experienced contractors and their legal accomplices. All three hypotheses are legitimate and, in concert with each other, are causing traumatic indigestion throughout the corps of highway agency administrators.

So again we call attention to the discipline of highway architecture coming to the rescue. One facet of the present system that must be corrected is the disjointed separation of phases in highway development. We have already addressed this a number of times. Nevertheless, by bridging these separations (planning, programming, location/preliminary design, final design, construction, operation, and maintenance), a much more cohesive fabric becomes available to highway agencies. It's also likely that contractors would get to appreciate this cohesion, reduce their reliance on court claims, and return to that which they do best: building roads and highways.

## CONSTRUCTION CONTRACTS

Although each agency formats and words its documents in its own distinct manner, there are many points in their makeup that run through all construction contracts. A typical contract is likely to include:

- Copy of a published notice or advertisement announcing the request for bids from the highway agency;
- The proposal submitted by the selected bidder, which lists the bid items, estimated quantities, price per unit, total contract price, firm name and address, together with principals of the firm;
- Errata sheet;
- Various agency requirements directed towards disadvantaged business enterprises, equal employment opportunities, affirmative action and nondiscrimination;
- Other agency requirements involving non-collusion, debarment, compliance, and certifications;
- Applicable environmental regulations dealing with air and water quality;
- Jingoistic boilerplates (e.g., *Buy America*);
- Listing of the minimum wage rates for various trades;
- Supplemental agency specifications that have been adopted subsequent to the last-published standard specification book;
- Special provisions that are unique to the project;
- Performance bonding certificate;
- Payment bonding certificate;
- Requisite methods of record keeping;
- Project plans; and
- Signature page.

## Cost Overruns

During the construction period and well into post-construction activities, the spectre of higher costs than were originally estimated for many contract items causes sleepless nights for highway agency employees.

Overruns on some items can sometimes be balanced against underruns on others—sometimes—to assure that the total contract price remains within bounds. What the architect must be alerted to are the other times when many more items overrun than underrun. These matters are usually negotiated (hopefully at the lowest appropriate administrative level, such as between agency field supervisor and contractor superintendent) at the time when it becomes evident that a significant quantity change in one or more items will be necessary. (Unexpected bad weather can do irreparable damage to a construction schedule—occasionally, in the northern states, extending contract completion into another construction season.) When it becomes necessary to modify contract provisions in the middle of a project, a change-in-work order or supplemental agreement is drafted and, if satisfactory to both parties, signed and executed. If not satisfactory to both parties, the adversary approach takes over and a previously established claims process or hearing gets rolling. Should the claims process not satisfy both parties, the matter has a good chance of winding up in the court system.

More often than not, the reason for overruns falls on the highway agency. It may be design errors, omissions on the plans, inadequate materials investigations, or poorly written specifications. Compounding these human errors, the agency structure may engender Balkanization: the breakup into small, noncooperating, competitive, and often hostile units within the agency. On the other hand, the inefficient contractor or the contracting firm that is operating out of its specialty may be worthy of blame.

On a country-wide basis, the specific most-often-cited factors contributing to overruns seem to be:

- Time constraints placed on the contractor that are too restrictive (some agencies use a combination of cost and time in order to determine who the actual low bidder is).
- Lengthy spells of bad weather that cause excessive time delays.
- Ambiguous contract provisions.
- Underground utility locations inadequately or incorrectly plotted or flagged.
- Design of a rehabilitation project relying too much on as-built plans from the last construction improvement on the site; subsequent maintenance activities might have changed highway characteristics and were not logged on the as-builts.
- Use by the contractor of wrong equipment for the job at hand.

In addition to these factors, there are actions that the alerted architect can take to reduce unnecessary cost overruns. Timely resolution of questions or disputes may do much to avert a contractor from resorting to the claims process. Having clear and accurate records also places the highway agency in a good negotiating posture. More complete sampling and testing of materials during the design phase may be beneficial, although a trade-off between the added cost of testing versus the anticipated benefits during construction should be considered. Finally, a postmortem on the completed project should be made, documented, and kept as a reference for future projects.

## RISK MANAGEMENT

The impact of cost overruns, together with immense awards in tort cases, has saddled highway agencies with a very significant financial burden—well above that needed to improve, maintain, and operate the national highway network. Response by highway agencies has been to set up risk management programs.

Sovereign immunity has been eroded over the past three decades so that in many jurisdictions it no longer exists to any degree. Smaller highway agencies, particularly at the local government level, are faced with high insurance fees to protect against future liabilities they may incur—parallel to the inordinate fees that certain medical doctors have to pay for malpractice insurance. Larger highway agencies have resorted to self-insurance, similar to the self-assurance that major railroad companies have employed through the years. This means that they must set aside large chunks of resource assets to cover anticipated court expenses and awards.

The highway architect may become involved in tort issues even though employees seldom are held liable for agency actions. Plaintiffs go after the "deep pockets." However, being named co-defendant in a $10 million lawsuit can be scary!

### References

Hinze, Jimmie, and James Couey. 1989. Time and weather provisions in construction contracts of state highway agencies. In *Transportation Research Record 1234*, pp. 57-63. Washington, D.C.: Transportation Research Board, National Research Council.

Lewis, Russell M. 1983. *NCHRP Synthesis 106, Practical Guidelines for Minimizing Tort Liability*. Washington, D.C.: Transportation Research Board, National Research Council.

NCHRP *Synthesis of Highway Practice 79, Contract Time Determination*. 1981. Washington, D.C.: Transportation Research Board, National Research Council.

Netherton, Ross D. 1983. *NCHRP Synthesis 105, Construction Contract Claims: Causes and Methods of Settlement*. Washington, D.C.: Transportation Research Board, National Research Council.

Sternbach, Jack. 1990. *NCHRP Synthesis 168, Contract Management Systems*. Washington, D.C.: Transportation Research Board, National Research Council.

Thurgood, Glen S., et al. 1989. *UDOT Highway Construction Contract Cost Overruns: Causal Factors and Remedial Actions*. Provo, Utah: Brigham Young University, Civil Engineering Department.

# 18
## Construction Items

A number of typical construction items were listed in Chapter 16 to summarize development of the design as outlined in preceding chapters. Since the step-by-step process in *designing* a highway is not quite the same as the step-by-step process in *building* a highway, the chronology of events following will reflect that difference. The one similarity between the two processes is that the steps overlap and intertwine. Because of the overlapping and intertwining endemic to the process, scheduling of the work (by the contractor) is a heavy and intricate responsibility.

### SCHEDULING

During the Second World War, the U.S. Navy developed a procedure, eventually to be known as the Critical Path Method (CPM), for scheduling its more complicated activities and missions. Subsequent to that time, many manufacturing and distribution firms adopted the CPM approach for scheduling. Highway contractors also jumped on the bandwagon. But CPM scheduling has never functioned as well in the highway field as it has in building construction. The main reasons for CPM scheduling failures in highway construction are that central office personnel rely on CPM too heavily while, at the same time, field personnel ignore it.

In the field, the more common scheduling technique is the bar chart. However, when visiting a jobsite, it becomes evident that the principal function of the bar chart is to show just how far behind schedule the job is.

Perhaps an effective means of scheduling a complicated project—one having many interrelated activities—is to attempt a mixed breed of CPM and bar chart. The advantages of jobsite pragmatics could be coupled with computerized CPM logic so that varying production rates, together with unanticipated delays, could be internally telescoped in order to maintain orderly sequences.

Although the highway architect may not be called upon to set up a project schedule, an understanding of the scheduling technique a contractor proposes to use is quite helpful when overseeing a construction project. Some of the most significant contract items, particularly those that are involved in scheduling, are described (more from the perspective seen in the construction phase) in the following sections.

## SITE PREPARATION ITEMS

Chapters 4 and 10 alluded to some of the things that must be accomplished before actual work can be undertaken. To begin, the survey party (or parties) establishes centerlines, offset lines, construction limits, reference points, bench marks (for elevations), and a host of other pins, nails, and stakes to witness what and where everything goes. Initiating a photo log of the site to record "before" conditions is a useful preventive measure to forestall potential disputes in the future.

### Clearing and Grubbing

Removing material readily visible on the ground surface, such as trees, stumps, logs, debris, and rubbish, falls under the heading of clearing. Grubbing consists of removing and properly disposing of roots, buried debris, and other obstructive materials not readily visible on the ground surface. Work of this nature ought not to go outside the bounds needed for actual construction operations. It usually includes selective thinning, rather than complete removal, of trees and woody vegetation. Tree trimming is also required in most standard clearing and grubbing specifications. A good practice is to return woody material back to the environment, either by disposition into the marketplace or by reducing it to chips and distributing it appropriately on the jobsite.

Special provisions relative to clearing the worksite may include developing scenic vistas via selective thinning, preparing vegetative fences for anticipated snow drifts, clearing to permit solar access on pavements in the wintertime, providing vegetation setbacks to reduce vehicle-animal collisions, creating greater site distances, and treating noxious and undesirable vegetation within the right-of-way.

### Salvaging Topsoil

Since topsoil has little value as foundation or fill material and high value for landscaping, the stripping and storing of topsoil takes place before, during, or immediately after clearing and grubbing. Topsoil should not be stripped if doing so could cause erosion problems. Blending lower quality topsoil with waste by-products or composted sewage sludge, if permitted by appropriate regulatory agencies, is useful as a recycling effort in promoting roadside landscaping.

### Obliterating Old Roads

There are occasions in the design phase when the removal, obliteration, and site restoration of abandoned road segments is overlooked, particularly when the new right-of-way limits do not include the old roadway prism. When this occurs and the contract documents do not allow for payment to obliterate old roads and restore the sites, the architect might wish to try some innovative methods to establish value for the old roadway materials or at least initiate a change in work order to get the oversight taken care of. Before doing so, a prudent inquiry into ownership of the old road segment can save a lot of embarrassment. Oftentimes, a state highway agency will transfer title to an old road to a local government agency. (What happens when the local government agency refuses to accept the gift?)

## EXCAVATION

Chapter 10 presented the designer with some of the basic parameters of dirt moving. On a construction project, the picture may unfold in a manner not fully expected. Site investigations are done in the design phase by sampling. When sampling or inadequate subsurface investigations do not show the quality of material actually excavated and when that quality of material actually excavated poses problems in extraction, transport, or deposition, other depositions (the kind handled by attorneys) take over and more business is assigned to the court system. The contractor likely has a legitimate and convincing case, based on information contained in the request for bids. The agency, in its own defense, may be able to contend that it made a sincere effort to obtain accurate subsurface information. If the issue can be resolved on-site, the agonizing litigation process can be preempted. So we are now led to the subject of negotiating skills required of every agency field supervisor whether he or she be project manager, resident engineer, or supervising highway architect. A master-level test of the supervisor's negotiating skills takes place when hazardous wastes are discovered during an otherwise routine excavation. Situations of similar nature continually crop up on the job and provide a challenge to the field supervisor, who cannot go crying to headquarters every time a problem comes up that is not detailed in the contract documents.

Excavation problems can be tricky, but potential embankment problems are able to be allayed most of the time. Placing the excavated material in fills can be checked, observed, tested, and controlled, if sufficient effort is directed to doing so. The architect must possess adequate knowledge and background in soil compaction, however, to achieve a modicum of success in this endeavor. This knowledge and background cannot be gained solely from the literature. Nonetheless, a few helpful hints are in order.

## Weather

Hot weather, cold weather, and wet weather all have detrimental potential for placing fills. Drying out of soil moisture, frozen chunks of dirt, and muddy conditions are circumstances in which common sense dictates that construction during bad weather is to be avoided. Standard agency specifications also contain many "don'ts." What the contractor needs to know is when conditions become marginally acceptable to begin or resume operations. The field supervisor's judgment is critical at this point.

## Equipment

Choice of equipment by a contractor for moving dirt is one of the more critical economic considerations. The right tool for the job is important. A front-end loader with a half-yard bucket attempting to move 100,000 cubic yards is equivalent to bailing out a sinking ship with a teaspoon. Likewise, to employ a series of 32-yard scrapers to dig a two-foot-wide drainage ditch is tantamount to swatting a fly with a pile driver. Terrain and anticipated weather conditions also play their part in selecting appropriate earth moving equipment. A crane with a long boom and dragline may have trouble remaining erect when working on sidehill locations; rubber-tired loaders can get mired in mud. And not to forget, operation of tall equipment in the vicinity of power lines is hazardous to the health and well-being of the operators. Another factor that trades off the scale advantages of large earth moving machines is the problem of getting these behemoths to and from the worksite. Should the highway architect view this as "the contractor's problem" ? Only if the architect is willing to abdicate some of the responsibility for assuring that a first-rate job is accomplished. If, however, a team approach (that is, an interdiscipline endeavor) is the intent, and in many cases it should be, the highway architect should be somewhat knowledgeable of the capabilities and uses of the more common earth moving tools. Such knowledge can be extremely handy in cases where an architectural design is applied to sculpturing the landscape adjacent to the highway. Using the correct equipment can do wonders toward obtaining the desired aesthetics.

Looking towards the future, application of the technology of robotics is quite likely to replace some of the time-proven techniques of excavating and grading. With appropriate sensing devices, robots can be effectively used where hazards and other uncertain conditions cause concern of injury to humans. Also, the scalar

Construction Items 225

**FIGURE 18-1.** Having the right tool for the job makes work a lot easier. (Photo: Colorado Department of Highways.)

magnitude of some earthwork components lends itself to the application of robotics.

Another aspect of equipment selection deals with soil compaction. Porous soils require vibration and pressure. Cohesive soils, such as dense clay, do not compact as easily and often need specialized equipment and procedures to accomplish satisfactory results. Where soil compaction is a large component of a project, the construction architect would find it worthwhile to consult local practitioners who are familiar with soil traits in the immediate area. Applicability of various roller types, vibratory compactors, segmented plates, and spreaders is really more of an art form than a scientific technology. More on the subject will be coming up in Chapter 20.

## Bridge Approaches

Touched upon briefly in Chapter 12, the dips in the pavement on each side of a bridge that become evident within a few years after opening to traffic can be averted if appropriate steps are taken in design and if meticulous attention is directed to the backfilling operation. Sometimes, lateral displacement of the soil in the deepest part of the fill (adjacent to the abutments) is cited as the cause should containment by wingwalls or retaining walls be omitted in the design. In such a case, the most meticulous attention during construction may be of little value.

## DRAINAGE AND EROSION CONTROL

Plentiful and varied are the items that a contractor may have bid on to provide adequate drainage and to protect the worksite from wind and water. Each different type of culvert or drain pipe has different properties and specifications, and they come in many sizes and shapes. Channel linings, splash pads, catch basins, drop inlets, leaching basins, headwalls, and similar appurtenances may be standardized, precast, or job-tailored. Erosion control may involve a great deal of measures and materials. Israelsen (1980), Appendix A, displays a significant array of techniques for controlling erosion, as well as locations and conditions where they are most effective.

Installing drainage and erosion control items in accordance with plans and specifications ought to satisfy these essentials at the time when construction operations are complete. Not so during the actual construction. It is in the contractor's and agency's best interest to assure that washouts and similar phenomena do not occur on the jobsite before all work is completed. Temporary drainage and erosion control must be continually on duty—even when all work is shut down for the winter. Many times, the contract documents will specify just what temporary measures must be included, but they can often be vague. Unless a highway architect is well versed in hydraulics, consultation with an independent authority in local hydrology/hydraulics (such as the hydraulics designer of the project) can verify whether or not the measures in use and proposed by the contractor will suffice.

## PAVEMENTS

From the design vantage, the subject of pavements was treated in Chapter 7. Here we'll go a bit further to assure that the design intent is fulfilled.

### Subgrade

No matter how sturdy the design and construction of the pavement itself, it will not last long if there is no support beneath it. Quality of the subgrade must be consistently high throughout the project. Points to be brought home in this regard are the potentials for inconsistency when:

- Various parts of the subgrade are prepared at different times or different seasons;
- Half the road prism is in cut and half is in fill;

**FIGURE 18-2.** Removing old PCC pavement by hydrodemolishing: a very high pressure water jet surgically cuts the concrete prior to the placing of a patch.

- Treatments such as soil-cement, PCC-treated-aggregates, or calcium chloride are applied to fortify segments of weak base material; or
- Adequate inspection and testing do not occur on a regular and continuing regimen.

Regarding the various field and laboratory tests, what is done if the subgrade does not test up to spec? Sometimes removal and replacing is necessary; at other times removal and replacing will do more harm than good. Most specifications permit a range of values in the testing. In situations where most of the continuing tests come within specs but at the lower range, it may be prudent to take measures that will assure that *all* tests come in at the lower range of the specifications. Those segments at the higher range portend some inconsistency. So the contractor bellows, "Hey, it's within spec, why do I have to (choose one of the following): remove and replace the good subgrade material/add water to the subgrade/dry out some of the soil moisture in the subgrade/change the type of compactor/treat with cement?" Response of the supervising highway architect must be couched in language to assure that the contractor is not going to lose money and that a better job will result. Although this scenario is oversimplified, it displays the type of actions required in many field situations.

## Portland Cement Concrete

Specifications for pouring Portland Cement Concrete (PCC) pavements are well defined by all major highway agencies. Some prefer reinforced, others non-reinforced. Thus, the construction supervisor is placed in the position of referee or umpire, to assure that all the rules are being followed.

Four aspects of PCC pavements that were treated in no depth at all back in Chapter 7 were joints, crack-and-seat, roller-compacting, and diamond grinding. They are dealt with here (and in Chapter 20) in a little more detail because construction techniques are more significant than design parameters.

Joints in PCC pavements are sawed-in shortly after the concrete has achieved its initial set. In most jurisdictions, the joint is then filled with a sealer, the purpose of which is to prevent moisture from entering the joint and working its way down to the subgrade. However, in some midwest and great plains states, the highway agencies have determined that these joint sealers are not cost-effective, so they saw thin joints and leave them unsealed. Effectiveness of the various materials and methods of joint sealing remains in the debating stage.

Crack-and-seat (along with its siblings, break-and-seat and rubblizing) is new enough to require the highway architect to keep up to date with its ever changing technology, along with evaluations of just how effective the procedure is over the longer term. Chapter 4 in *NCHRP Synthesis 144* provides a concise primer on crack-and-seat construction. It appears that this procedure has a good deal of promise.

Roller-compacted concrete has been used primarily for heavy-duty transfer aprons, large parking lots, and other locations that do not require a smooth riding surface for high-speed vehicles. It does away with slip forms and employs a paver with vibrating screed and tamper. Compacting is then accomplished in three steps: vibrating smooth-wheel roller, pneumatic roller, and finally a non-vibrating steel drum roller.

Diamond grinding is applied to a rough PCC pavement to restore rideability. The pavement must otherwise be in good shape (i.e., minimal cracks, spalling, and joint failures). Darter (1990) presents some findings on how effective diamond grinding has shown to be.

## Asphalt

In the same manner as with PCC pavements, the field supervisor acts as arbiter to assure that the plans and specifications are being adhered to. Testing throughout necessarily tends to be more extensive because of the plastic nature of asphalt pavements: applied forces cause them to creep (as do most geo-textiles used in conjunction with asphalt pavements).

Joints are rarely placed in flexible pavements. But when an asphalt overlay is placed atop a PCC base, joints are usually sawed directly above the PCC joints so that expansions and contractions from the PCC slabs reflecting upwards do not appear as cracks in the surface.

Recycling of old asphaltic pavement materials began in earnest during the 1970s. Two approaches are popular: hot and cold (neglecting surface heater-planer reconditioning). In both processes, the old materials are blended with new before being placed back as a base course. A wearing course exclusively of new material tops off the surface. Photographs in Chapter 2 of Epps (1990) show the various operations involved in recycling materials from old pavements.

The nature of asphalt pavements requires that they be tended to on a regular basis. For this reason, Chapter 22, dealing with maintenance, will chronicle more tidbits that may sometimes be applicable to construction and reconstruction.

## STRUCTURES

More items are contained under this topic than can herein be addressed to any length. Unless schooled in most of the fine points of structural engineering and foundations, the construction supervisor is well advised to obtain expert assistance when dealing with bridges and other complex structural components included in the highway contract.

## ROADSIDE

Unlike structure items, roadside components tend to be the most interesting for the highway architect who takes pride in a top-notch effort. Even including those facets of roadside in which the highway architect has little or no background, a good and earnest effort in dealing with roadside items during the construction phase can

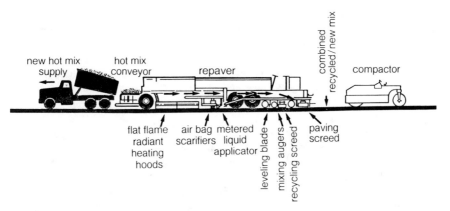

FIGURE 18-3. Recycling older bituminous pavements is gaining acceptability. (Sketch: Cutler Repaving Inc.)

mean the difference between a so-so highway and one that can be considered a masterpiece. (Perhaps that's overdoing it a little, but you get the idea.)

Aside from the fun stuff, utility relocation and protection can be the largest set of items that fit under the umbrella of roadside. They can be quite the migraines, but they require dutiful attention. If utility issues are resolved early on in the construction phase and if they are handled timely and routinely, the results can add in a positive way to a harmonious project completion.

**References**

AASHTO. *A Guide for Transportation Landscape and Environmental Design.* 1991. Washington, D.C.: American Association of State Highway and Transportation Officials.

Darter, Michael I., and Kathleen T. Hall. 1990. Performance of Diamond Grinding. *Transportation Research Record 1268*: 43-51.

Epps, Jon A. 1990. *NCHRP Synthesis 160, Cold-Recycled Bituminous Concrete Using Bituminous Materials.* Washington, D.C.: Transportation Research Board, National Research Council.

FHWA. *Improved Rigid Pavement Joints, Report RD-86/040.* 1986. McLean, Virginia: Federal Highway Administration.

Friend, David, and Jan Connery. 1988. *NCHRP Report 310, Dealing With Hazardous Waste Sites.* Washington, D.C.: Transportation Research Board, National Research Council.

Israelsen, C. E., et al. 1980. *NCHRP Report 221, Erosion Control During Highway Construction.* Washington, D.C.: Transportation Research Board, National Research Council.

Kilareski, Walter, P., and Richard A. Bionda. 1990. Sawing and sealing of joints in asphaltic concrete overlays. *Transportation Research Record 1268*:34-42.

Mitchell, James K., and Willem C.B. Villet. 1987. *NCHRP Report 290, Reinforcement of Earth Slopes and Embankments.* Washington, D.C.: Transportation Research Board, National Research Council.

Najafi, Fazil T., and Sumath M. Naik. 1989. Potential Applications of Robotics in Transportation Engineering. *Transportation Research Record 1234*:64-72.

NCHRP *Synthesis of Highway Practice 47, Effect of Weather on Highway Construction.* 1978. Washington, D.C.: Transportation Research Board, National Research Council.

NCHRP *Synthesis of Highway Practice 115, Reducing Construction Conflicts Between Highways and Utilities.* 1984. Washington, D.C.: Transportation Research Board, National Research Council.

Robinson, Gary K. 1991. Project constructibility. Paper read at the annual meeting of WASHTO, July 12-18, 1991, Belleview, Washington.

Scherocman, James. 1991. *Hot-Mix Asphalt Paving Handbook.* Lanham, MD: National Asphalt Pavement Association.

Skibniewski, M. J. 1988. *Robotics in Civil Engineering.* New York: Van Nostrand Reinhold.

Thompson, Marshall R. 1989. *NCHRP Synthesis 144, Breaking/Cracking and Seating Concrete Pavements.* Washington, D.C.: Transportation Research Board, National Research Council.

Transportation Research Record 1282. 1990. *Transportation Construction*. Washington, D.C.: Transportation Research Board, National Research Council.

Wahls, Harvey E. 1990. *NCHRP Synthesis 159, Design and Construction of Bridge Approaches*. Washington, D.C.: Transportation Research Board, National Research Council.

Wright, Paul H., and Radnor J. Paquette. 1987. *Highway Engineering*. New York: John Wiley & Sons.

# 19
## Safety

Throughout preceding parts, in those chapters dealing with preliminary and final design, the subject of safety was continually alluded to, culminating in a safety review outlined under "Plans" in Chapter 16. The actual application of myriad safety precautions begins, however, when the project is put under construction. Safety measures continue to hold a paramount position through completion of the project, opening the highway to traffic, and during all stages of operation. Safety directly affects human life so that it overrides all other considerations that have an influence on work of the highway architect.

### WORK ZONE SAFETY

A startling increase in work zone accidents has come to the attention of the highway community in recent years. Noncompliance by vehicle operators appears to be a major factor in the increase. Disrespect for traffic controls and for speed limits that are perceived to be too low has been found to be particularly prevalent in geographic belts where warning sign removal and obliteration of no-longer-applicable pavement markings are delayed or neglected. The architect, therefore, ought to be aware of just what practices prevail in the area and apply appropriate caveats. Some help can be found in the literature. Removal of old pavement markings is addressed in NCHRP *Synthesis 138*, pages 9–12. Upchurch, *et al.* (1992) evaluates variable message signs. Seventeen case studies of traffic management during construction are documented in the FHWA publication, *Traffic Management During Major Highway Reconstruction*. In places where construction impacts would completely tie up normal daytime traffic on a heavily traveled road and a detour or shoo-fly is not feasible, the alternative of restricting construction to nighttime hours might be considered. *TRR 1086*, pages 31–36, offers guidance on this elective.

Traffic at most worksites is rarely more than 50 feet from construction activity; occasionally, it comes as close as 1 foot from the work going on. This leaves very little room for error—both on the part of the driver and on the part of the worker. When vehicles go speeding by, they can create whirlwinds of dust and loose objects. Together with the noise they generate, autos, trucks, and buses foster an extremely unsafe workplace.

Part of the problem, paradoxically, is that modern vehicles provide such comfort to the occupants that they tend to be unaware of the environment immediately outside. One has just to compare the turbulence generated from the zipping by of an auto traveling 55 MPH to the interior ambience of the vehicle while driving at the same speed. Motorists perceive little threat to their own lives and act accordingly in directing inadvertent threats to others. The hub of efforts to reduce work zone accidents is to keep the speed of passing traffic within appropriate bounds.

**Slowing the Traffic**

Although many construction personnel blame motorists for being [expletive deleted] ignorant, many times the fault can be placed on the construction activity and certain overdone concomitant measures to assure safety: the "cry wolf syndrome." As mentioned, disrespect for signs and delineation that do not apply is a normal human reaction (e.g., a "slow—work zone ahead" sign that remains uncovered after all work has terminated for the day). Where such careless concern prevails, it follows that any amount of signing and barricading will not attain optimum effectiveness. Should vehicle operators become jaded or inured by too much signing and too many warnings, there are other techniques that can be used to slow the traffic entering a construction zone. They form two categories:

1. Legal and accepted practices that have become standard and sometimes work; and
2. Nonstandard practices that are frowned upon but that usually work.

The two most efficacious techniques in the first category are flagging and law enforcement. Speed reductions of 15 to 20 percent can be attained. These two techniques can become expensive in situations where construction activities are intermittent, as on a long stretch of highway undergoing spot improvements. Also, it's necessary to determine if the local law enforcement agencies have the capability and willingness to support their share of the endeavor.

In the second category, we come to the questionable practices that are likely to bear a great deal of criticism from many sources. If the construction site is not readily visible to oncoming traffic as, for instance, over the crest of a hill, a minor distraction on the approach to the site—keeping in mind that the distraction must not avert the driver's attention away from the road and other traffic—can cause the operator to remove the foot from the accelerator pedal. The distraction might be the sight of a simulated accident where an old junked auto is rolled over on its side some 30 feet or so from the pavement. If this is not acceptable, the positioning of a large piece of construction equipment at the same spot can aid in reducing the speed of passing vehicles. Should this not be practical, placing a vehicle equipped with flashing blue-and-reds (and radar gun) just off the shoulder will at least get the passing motorists' attention to the extent that they will slow down close to the posted speed limit. (Acquiescence of the local police is mandatory.) As with too many signs, these "cute" tactics diminish in effectiveness as drivers get wise and start to ignore them; they have to be varied on a regular basis or else they lose all usefulness.

Somewhere between the accepted practices and those that may be unacceptable are methods for unique situations. Example: On a stretch of road where visibility is normally no problem, large screaming signs, flashing barricades, cones and channelizing tend to obscure the actual work in progress. What often functions better is to remove the excessive warning and channeling paraphernalia and replace them with a small sign on the approach to the construction activity. Vehicle operators react much more positively when they can actually see work activity. Also, the smaller sign tends to cause vehicles to slow down so that the drivers can read the message. Volume 4, *Speed Control in Work Zones,* of the FHWA publication *Improvements and New Concepts for Traffic Control in Work Zones* outlines most of the standard and some of the innovative attempts to reduce speed through work zones.

WARNING: In locales where tort liabilities are the preeminent lifestyle, use of any of the above-suggested methods that do not conform to MUTCD and agency standards is sure to travel the serpentine paths through jurisprudence devices. (The lawyers will eat you up.)

### Protecting the Workers

Few of the barricades used to channel traffic can withstand much impact by an errant or out-of-control vehicle. Construction workers need additional protection. Truck-mounted flashing signs can provide a certain amount of shelter if they are located between approaching traffic and the worksite. When closing off a lane for a short distance, such as for a bridge repair, many contractors opt for "portable" precast jersey (New Jersey) barriers. When autos hit them at high speeds, they tend to bounce off; when higher trucks and semis hit them, there is the danger of rollovers and resulting carnage.

It may not be the full responsibility of the highway architect to assure worker safety, but any time it appears that an accident is waiting to happen, the architect ought to bring it to the attention of those who are accountable.

## Maintenance and Protection of Traffic During Construction

On most rebuilding jobs, the potential nuisance of (and to) traffic during construction initially has to be tackled during the design phase. Highway agencies have developed, over the years, policies, and guidelines to handle maintenance and protection of traffic during construction. Sometimes, the responsibility lies entirely with the construction contractor who provides the agency field representative with a Traffic Control Plan (TCP) for approval prior to commencing construction. Occasionally, an agency will include a rigid traffic scheme in the contract documents, allowing no flexibility to the contractor. Most projects, though, fall between these two extremes.

As it generally happens, the designer, preferably in consultation with field engineers, puts together a preliminary or recommended traffic control scheme consisting of plan sheets and phasing instructions to the extent necessary to assure both that the project can be constructed expeditiously and that anticipated vehicular and pedestrian traffic will be handled in a safe manner. Projected traffic counts determine approximately what extent, size, and type of temporary bypasses (shooflies), detours, lane restrictions, access openings, flagging, pavement markings, and so on will be needed to facilitate the anticipated traffic (albeit almost always at a lower level of service) during the period of construction. An architect may well be alerted to many changing conditions because, in spite of all the policies, standards, and guidelines that have been promulgated, maintenance and protection of traffic during construction remains an activity requiring a good deal of judgment applied to each individual project.

In all cases, consistency with driver expectation is the best measure to apply when reviewing a scheme for work zone safety and traffic accommodation during the period when the roadway improvement is being built. Refer to the *Manual on*

**FIGURE 19-1a.** Work zone protection (or lack thereof) varies considerably from one jurisdiction to another.

236　Highways: An Architectural Approach

**FIGURE 19-1b, c, d.** Work zone protection (or lack thereof) varies considerably from one jurisdiction to another.

*Uniform Traffic Control Devices*, augmented by its little brother, *Traffic Control Devices Handbook*, applied to the directives of *FHPM 6-4-2-12(6a)*, for the best way to initiate maintenance and protection of traffic during construction.

## SAFETY AFTER THE FACILITY IS OPENED TO TRAFFIC

More on this topic will be forthcoming in Chapter 21. Yet there are a few thoughts for the highway architect to keep in mind during the construction phase. Safety-related particulars and objects that can be difficult to relocate once the contractor leaves the scene ought to have all doubts and issues resolved before a highway is opened to traffic.

### Guard Rails

In some jurisdictions, the term is *guide rails*. Their type, makeup, and placement are well outlined in agency standards. Likewise with median barriers. Yet when something appears to be out of place—even if conforming to standard—it's incumbent on the field supervisor (or inspector) to determine what corrective action should and can be taken. One common example would be the conflict between a guard rail and utility installation where standards of two different agencies direct that both pieces of highway furniture occupy the same space. Another illustration would be where "strong" guard rail posts are specified, yet a safer condition may be attained at specific locations by utilizing the resilient properties of "weak" posts.

### Utility Poles

Breakaway poles are used to save the lives of occupants when a vehicle veers off the beaten path and tries to wrap itself around a roadside object. But what if the pole is supporting an overhead high tension electric line? The danger of live wires tumbling into traffic is an even greater hazard. All efforts to remove utility poles from shoulder and roadside locations should be employed, including poles and similar obstacles that may be encountered by an errant vehicle traveling down a flat slope outside of the clear zone.

### Truck Escape Ramps

An errant vehicle traveling down a flat slope poses a danger to the vehicle occupants; an errant vehicle, such as a runaway truck which has lost its brakes, traveling downhill on the pavement poses danger to all traffic in the vicinity. Proper placement of one or more truck escape ramps on long downhill highway segments is relatively important. On two-lane highways, the runaway truck ramp must be located off the right side; however, on divided highways, if possible, the ramp should be off the left (median) side because the brakeless truck or semi is probably traveling well over the speed limit and should not be encouraged to remain in the "slow" lane.

Construction of the ramp itself can increase its effectiveness. Most ramps use gravel for the arrester bed. It works, but many truckers who have had to use escape ramps dread the prospect of ever using one again. They may attempt to "ride it out" rather than experience a second time the dreadful "g" forces of the sudden deceleration and stop. A simple means to avoid the excessive deceleration force is to gradually increase the depth of gravel at the entry of the escape ramp rather than to build it full depth for its entire length. (Most trucks need only the first third of a properly designed ramp to come to a complete stop.)

## Pitfalls in Constructing the Safe Highway

For every attempt to do the right thing, there seem to be at least two reasons to do just the opposite. Removing one hazardous condition might in turn create several new hazardous locations. Does this mean that the hazardous condition must remain as is? Certainly not. The point to realize is that a great deal of thought and care is required to do the job properly—to avoid introducing impairments as components of improvements. Sometimes an alleged "safety measure" becomes a two-edged sword. As an example: the safer the grate over a drainage basin, the lower the hydraulic efficiency—which, in turn, requires a larger size grate—which, in turn, further impairs safety. Another "for instance": guard rails are installed essentially to protect against roadside hazards; paradoxically, guard rails in themselves are also roadside hazards. There can be times when a weary highway architect becomes infected with the "You can't win" syndrome.

Advice: Take a break and come back with more "fire in the belly" !

### References

AASHTO. *Guide for Selecting, Locating and Designing Traffic Barriers.* 1977. Washington, D.C.: American Association of State Highway and Transportation Officials.

Bowman, B. L., J. J. Fruin, and C. V. Zegeer. 1989. *Handbook on Planning, Design, and Maintenance of Pedestrian Facilities.* FHWA-IP-88-019. McLean, Virginia: Federal Highway Administration.

FHPM 6-4-2-12, *Traffic Safety in Highway and Street Work Zones.* Washington, D.C.: Federal Highway Administration.

FHWA. *Handbook of Highway Safety Design and Operating Practices.* 1978. Washington, D.C.: Federal Highway Administration.

FHWA. *Improvements and New Concepts for Traffic Control in Work Zones,* 1985. McLean, Virginia: Federal Highway Administration.

FHWA. *Manual on Uniform Traffic Control Devices.* Regularly updated. Washington, D.C.: Federal Highway Administration.
FHWA. *Traffic Control Devices Handbook.* Regularly updated. Washington, D.C.: Federal Highway Administration.
FHWA. *Traffic Management During Major Highway Reconstruction: Abbreviated Case Studies.* 1988. Washington, D.C.: Federal Highway Administration.
King, G. F. 1989. *NCHRP Report 324, Evaluation of Safety Roadside Rest Areas.* Washington, D.C.: Transportation Research Board, National Research Council.
NCHRP Legal Research Digest 6. 1989. *Impact of the Discretionary Function Exception on Tort Liability of State Highway Departments.* Washington, D.C.: Transportation Research Board, National Research Council.
NCHRP Report 236. 1981. *Evaluation of Traffic Controls for Highway Work Zones.* Washington, D.C.: Transportation Research Board, National Research Council.
NCHRP Synthesis 128. 1986. *Methods for Identifying Hazardous Highway Elements.* Washington, D.C.: Transportation Research Board, National Research Council.
NCHRP Synthesis 132. 1987. *System-wide Safety Improvements: An Approach to Safety Consistency.* Washington, D.C.: Transportation Research Board, National Research Council.
NCHRP Synthesis 138. 1988. *Pavement Markings: Materials and Application for Extended Service Life.* Washington, D.C.: Transportation Research Board, National Research Council.
Transportation Research Record 1086. 1986. *Roadway Markings and Traffic Control in Work Zones.* Washington, D.C.: Transportation Research Board, National Research Council.
Transportation Research Record 1127. 1987. *Innovation, Winter Maintenance, and Roadside Management.* Washington, D.C.: Transportation Research Board, National Research Council.
Transportation Research Record 1133. 1987. *Roadside Safety Features.* Washington, D.C.: Transportation Research Board, National Research Council.
Transportation Research Record 1163. 1988. *Traffic Control in Work Zones.* Washington, D.C.: Transportation Research Board, National Research Council.
Transportation Research Record 1230. 1989. *Work-zone Traffic Control and Tests of Delineation Material.* Washington, D.C.: Transportation Research Board, National Research Council.
Transportation Research Record 1233. 1989. *Design and Testing of Roadside Safety Devices.* Washington, D.C.: Transportation Research Board, National Research Council.
TRB Special Report 218. 1988. *Transportation In An Aging Society.* Washington, D.C.: Transportation Research Board, National Research Council.
TRB State of the Art Report 6. 1987. *Relationship Between Safety and Key Highway Features.* Washington, D.C.: Transportation Research Board, National Research Council.
Upchurch, Jonathan, *et al.* 1992. Evaluation of Variable Message Signs: Target Value, Legibility, and Viewing Comfort. Paper read at 71st Annual Meeting of the Transportation Research Board, January 12-16, 1992. Washington, D.C.

# 20
# Inspection

Rule number one in construction inspection is that quality of inspection outranks quantity of inspection. A following corollary is that intermittent inspection by a qualified, motivated inspector is superior to full-time inspection by an individual who doesn't know (or care) what or how to inspect.

Rule number two states that most contractors (who depend on continued work from highway agencies) will perform in a conscientious manner to provide a good end product. They do not appreciate, neither do they respond well to, being treated as convicts on a chain gang. Inspection techniques need to take a cooperative rather than adversary approach (but not too cooperative).

## STAFFING FOR A CONSTRUCTION PROJECT

What specialties and how many of each discipline should be represented to inspect work on a construction project? No precise or definitive answer can be made to such a general question. More must be known. The nature of the project (i.e., clear and excavate, grade and drain, pave, build bridges, landscape, install signs, signals, and lighting, or any combination thereof) has a lot to do with selection of disciplines. Size or scale of a project (i.e., $100,000, $1,000,000, $10,000,000, $100,000,000) influences how many agency representatives are to be assigned to the field crew. There are other determiners, such as geographic region, reputation of the contractor, and, above all, agency policy. (Variations in state highway agency staffing of projects is described in *NCHRP Synthesis 145*.)

Even on the smallest projects, there are several basic tasks that must be performed:

- Locating control points and limits of work;
- Assuring that safety precautions are adequate;
- Inspecting work;
- Testing (or certification) of materials;
- Maintaining accurate records;
- Preparing estimates of (monthly) progress payments due the contractor; and
- Responding in timely fashion to questions regarding interpretation of plans and specs.

When only one or two persons are assigned to a field crew, the mastering of a single specialty is of less importance than a smattering of knowledge on a wide range of topics. For this reason, some agencies assign their most experienced personnel to the smaller projects. Other agencies use the small projects as training grounds for the neophytes and less experienced.

Gargantuan projects are likely to have a complement that includes a full office staff, two or three survey parties, technicians operating on-site testing laboratories, computer programmer, structural engineer, geologist, archeologist, hydrologist, public relation specialist, legal intern, inspectors schooled and experienced in a multitude of disciplines, landscape architect, deputy project engineers, project manager (highway architect), and/or a highly credentialed resident engineer.

Whatever the staffing, there are some generalized approaches to inspection of contract work that apply universally. The sections and subsections that follow are arranged as in Chapter 18. They represent only a small sample of items encountered on a major construction project, but provide a microcosm from which the highway architect can expand as experience and opportunities present themselves.

## SITE PREPARATION

Inspectors often find themselves conscripted into survey parties at the beginning of work on a project. This is good because, if nothing else, the inspector gains a knowledge first hand of what the project limits are and where many of the control points are located.

### Clearing and Grubbing

Where payment for some work (e.g., undercutting unstable material) is by volume, the inspector should assure that preliminary cross sections (or other survey measurements) are taken prior to removal of material. In fact, anything that requires before and after documentation must be logged in before work commences. This is also the best time for the inspector to set up a field book/diary rather than waiting for the first concrete pour.

A problem often cited by environmentalists is the burning of vegetation by contractors after they have cleared their worksites. If there is to be any burning, the inspector should make sure that the contractor has obtained the requisite permits to do so and has fires properly tended through the burn period.

**Topsoil**

If topsoil is to remain on-site through the duration of the contract—possibly through two or three construction seasons—it must be adequately nurtured, drained, and protected from erosion. Good topsoil generates lots of vegetation when in storage, but before it is spread during the landscaping phase, the vegetation (weeds) must be prevented from reseeding.

Occasionally, a contractor, reacting to market pressures, will sell off some of the higher quality stripped topsoil before it is stockpiled on-site, and replace it with soil of a lesser quality. Inspectors need to be alert to this temptation.

**Obliterating Old Roads**

Common sense plays the domineering role in decisions regarding how obliterated an old road should become. The finished product ought to fit in with the surroundings. Proper clean-up is essential.

## EXCAVATION, BACKFILL, AND EMBANKMENT

During the first three decades of interstate construction, more inspection time was probably spent on earthwork than on any other phase of highway work. Now that major earth moving forms a lesser part in highway reconstruction, the person-hours spent inspecting will be less, but quality of inspection must remain high and thorough. A good basic reference on earthwork construction is the TRB *State of the Art Report 8*.

**Compaction**

Compacting trenches, particularly the ones that cross transversely (90 degrees) under the proposed pavement, must be controlled and checked almost to the same degree that a diamond cutter exercises when trimming a 20-carat stone. Although the specifications probably call out many provisions relating to method of tamping and thickness of backfill layers and quality of backfill material, the inspector on the job must have a good feel for the adequacy of the work performed.

Use of the correct compaction equipment enhances the prospects of achieving best results. In narrow and hard-to-reach places, tampers are more useful than rollers. The latter come into play when larger masses of fill are to be compacted,

**FIGURE 20-1.** Single-drum smooth vibratory compactors are used for granular and semi-cohesive soils. The larger models are also used to compact soils heavy in gravel or rock. (Photo: Caterpillar Paving Products Inc.)

but each soil classification responds best to a specific type or types of roller. Rock fills and gravel seem to be the most difficult to compact properly. Rubber tires in combination with a smooth vibratory drum work best. Sandy and silty soils can be compacted with a smooth drum vibratory roller. When considerable clay is encountered, a padded drum (sometimes referred to as sheepsfoot) obtains the best compaction. Unique properties of local soils may necessitate that nontraditional or special compacting equipment and methods be used.

Method of compaction can and should be subject to an inspector's judgment. Visual observations of how well compactors and earth moving equipment behave when traveling back and forth over successive layers of fill material is a pivotal reason for having an inspector on-site. The inspector can also see to it that hauling and leveling equipment is routed over the entire width of an embankment rather than being confined to one path.

Where excavating for structure footings that must support great weights, it is important for the contractor to excavate just to the elevation of the bottom of the footing so that it is not necessary to bring fill back in (presuming that the consolidated soil is stable and has the essential strength). The fill is not likely to have the same bearing capacity as the undisturbed earth. (Maybe that contractor at Pisa overexcavated on one side, below the footing for the tower, and then, when the inspector was out to lunch, backfilled the overexcavated portion with boggy organic material.)

244    Highways: An Architectural Approach

**FIGURE 20-2.** Pneumatic rubber tire rollers are employed both on granular soils and asphalt pavements. (Photo: Bomag U.S.A., Inc.)

At the completion of grading operations, tidying up helps to achieve a smooth landform transition. Rounding slopes at the tops and bottoms and removing machine-made berms (which tend to show up at the limits of work) enhances the aesthetics and contributes in some measure to erosion control. Contractor staging areas and borrow sites should be subjected to the same tidying up standards as the construction site itself.

**Testing**

Considerable testing of the soils was probably done during design of the project, yet the ultimate proof of the successful design is contained in the completed project. Therefore, during construction operations, continual testing is prudent at the least—more so when the sponsoring agency deems it necessary.

Density of the soil below a pavement subgrade or a structure foundation is critical. Testing for density in the field is now done by nuclear gauges more often than by time-consuming laboratory methods. The laboratory is relegated to the task

**FIGURE 20-3.** Double-drum smooth vibratory rollers are also utilized on granular soils and asphalt pavements as well as semi-cohesive soils. (Photo: Caterpillar Paving Products Inc.)

of determining the causal factors when the nuclear gauge readings do not show adequate density. Although non-optimum moisture content is the culprit most of the time, other soil characteristics, such as improper gradation or excessive organic material, may contribute to low-density readings.

Under bridge footings, soil strength is another important characteristic. Laboratory testing is performed on undisturbed field samples to assure that the load carrying capacity of the soil is equal to the task.

## DRAINAGE AND EROSION

Natural erosion is vital to sustain healthy waterways and is not considered pollution. By contrast, excessive erosion, which often occurs on larger earthwork undertakings—running up to one thousand times the natural erosion—is the single greatest pollutant by volume in the nation's waterways.

### Drainage and Culvert Pipes

Inspection of pipelines includes certification that the size, class, type, and manufacture of each pipe length is in accordance with the specifications. Before pipe laying, the bed must be checked for proper grade, compaction, and shape. Tongues of each

**FIGURE 20-4.** Padded drum compactors are best for cohesive soils—those high in clay content. (Photo: Bomag U.S.A., Inc.)

pipe length should be in the direction of flow, which means that pipe laying is to proceed from outlet to inlet (working upstream). Joints are to be snug and the pipe aligned as straight as possible—particularly vertically so that there are no dips that could catch and hold stagnant water. Invert (flowline) elevations must be checked at every junction of drainage pipes with drainage structures and at each end of all culverts. Backfilling pipelines is ticklish because tamping of each layer must be vigorous enough to consolidate the soil, yet gentle enough so that the operation does not damage the pipe beneath. The above-cited inspection mandates are usually included in agency standard specifications, but can be overlooked when the least experienced inspector has not been afforded access to the standard spec book and is sent out to check the drainage work.

Near the time of completion for all contract work, culvert and drainage pipes should be checked (from inside) to make sure that they are clean, clear of obstructions, and undamaged and that the flowlines have not been altered by settlement, backfilling, or other nearby construction activities.

### Drainage Structures

Some of the items in this category come precast, some must be constructed on the jobsite, and some built in-place. Plans, specifications, and special provisions outline most of the inspector's responsibilities. It's up to the inspector's supervisor to make sure that the inspector has access to the appropriate documents.

**Erosion Control Measures**

This portion of the work requires continued diligence. Changing jobsite conditions require that timely placement of both temporary and permanent erosion controls is undertaken. Since temporary measures are mostly a contractor's option, unless otherwise stated, the inspector's primary responsibility is to take notes of what temporary controls are implemented and when the contractor installs them.

**Geo-textiles**

If fabrics are able to withstand the stresses of being placed (and having construction machinery run over them), they will probably hold up quite well during their expected life. There is one caution, however: ultraviolet light. On slopes facing north, the problem is minimal. Otherwise, direct sunlight can soon destroy exposed geo-textile materials. Lapping of fabrics, as common sense would tell, should be performed so that water flows over, not into, the laps. Since the trend towards use of geo-textiles in highway applications is fairly recent, problems have arisen from inappropriate materials being specified. (A harried designer in a rush to meet a project advertising deadline may have discovered a geo-textile type used for a pavement interlayer from a previous project and specified its use for slope protection.) Woven and nonwoven materials are not interchangeable. Other geo-textile attributes include tensile strength, puncture strength, burst strength, permeability, elongation, and resistance to abrasion. If it is possible to certify the correctness and applicability of these various properties before and during placement, all the better. When this is not feasible, the best test is to examine how well the material stands up while being placed and covered (and repaired).

# PAVEMENTS

Some inspectors make an entire career out of pavement inspection, suggesting that perhaps technical education is not as important as diligence and experience. An old hand at pavement inspection can recite (to a young hand) a litany of cases where excellent results were achieved when everything on the job went wrong, followed by a chronicle of early pavement failures when all the specifications were met. More often than not, considerable elaboration accompanies these recitals. The well-established fact is that, when specifications are followed, the results turn out much better than when they are shortchanged. Exceptions do occur, but they are so rare that they stand out and are remembered as fondly as bursitis pains in the experience of the old timers.

## Subgrade

As we've mentioned before, no pavement will stand up for long if the subgrade collapses. Inspection of the subgrade immediately before placing of the subbase or base course or concrete slab is the most critical time to do so. Such advice presumes that preparation of the subgrade was watched over properly. Standard agency specifications should suffice in outlining an inspector's duties.

However, with crack/break-and-seat and rubblized base courses, the subgrade is not visible. Few agencies have long-standing standard specifications for these techniques. Certain initiatives may have to be exercised by the inspector to assure that the subgrade is, in fact, adequate and consistent. Having heavy equipment (not the compactors) run back-and-forth over questionable areas, watching closely as the wheels pass to observe any noticeable deflections in the broken concrete may be one method of assurance.

## Portland Cement Concrete

Just before placing PCC on a subgrade, if that is the pavement design, the subgrade is usually dampened so that the subgrade soil does not draw moisture out of the concrete, which, presumably, is at the optimum water content to properly hydrate the cement. Besides observing the operation, verifying adequate mixing times, assuring that line and grade are correct, recording deliveries, and reminding the construction personnel that PCC pavements are to be poured and finished as a continuing non-interrupted operation, the inspector may also be required to take air-entrainment readings, slump tests, and prepare cylinders. It's helpful if an inspector has the correct equipment to perform these on-site activities and a knowledge of the correct ASTM and/or AASHTO procedures. An inspector can be kept quite busy during a pavement operation.

Inspection at the batch plant can get to be a drag. An empathetic project manager (who may well be the highway architect) will see to it that batch plant inspection personnel is rotated.

The two most essential components in a PCC mix are:

- Adequate cement; and
- Not too much water.

Checking these are two of the inspector's prime responsibilities. The other responsibility—usually assigned to the most conscientious and experienced inspector—is proper finishing of the pavement surface. Timing of texturing, timing of joint sawing, and application of curing compound are very sensitive: narrow windows exist between too early and too late, or between too much and too little. Both temperature and humidity are important factors in determining the optimum times to perform the finishing touches. Specifications for concrete finishing tend to leave a lot to the inspector's best judgment.

Roller-compacted PCC pavements will not achieve the rideability of traditionally poured pavements. But even to get a fairly good surface, the inspector has to assert a fair degree of fussiness.

## Asphalt

Construction projects rely on plant mix bituminous concrete or a combination of plant mix and recycled materials. Thus, inspection must also take place both at the batch plant and at the laydown machine. Although continuity of operation is important, it is not as critical as with PCC pavements. The things that the inspector must watch out for are:

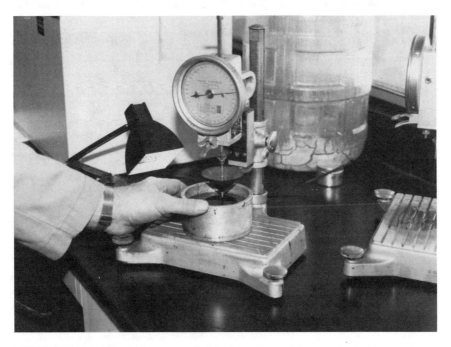

**FIGURE 20-5.** A traditional laboratory method for determining consistency of asphalt is to measure the speed of penetration by a standard needle (AASHTO T49).

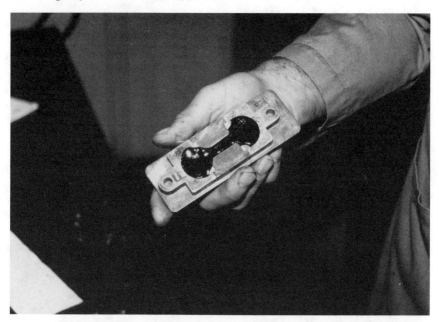

**FIGURE 20-6a and b.** Asphalt ductility is measured by stretching (without breaking) a 77°F briquet specimen in a mold at a speed of 50 millimeters per minute in a water bath (AASHTO T51).

- Correct percentage of bituminous material (too much will bleed, too little will not provide adequate bonding);
- Proper temperature at delivery and laydown;
- Thorough tacking (where required) to the course below; and
- Careful rolling and compacting (the drive wheels of the rollers are usually required to be nearest the paver).

## STRUCTURES

A very basic, yet sometimes overlooked, item is to make sure that a bridge or other major structure (including foundation piles for same) is laid out in the correct place and that nearby survey bench marks have been checked for accuracy. It can be embarrassing if a formidable highway bridge has to be jacked up three feet or, worse, moved several hundred feet after it is half completed.

Bridge inspection is somewhat of a specialty, but there are a number of generic features that can easily be checked by a journey highway inspector. Formwork, reinforcing bars, any observable damage to beams and girders that may have occurred

**FIGURE 20-7.** Recovery of asphalt from a field sample—to determine the percentage of asphalt in the mix—is done by boiling down and distilling via the Abson Method (AASHTO T170).

during transit or unloading, concrete placement, and normal testing procedures all lend themselves to complying with plans, standard specifications, and common sense.

## ROADSIDE

There would be little point to go into the hundred or more inspection attributes that could arise from the various roadside items that may be included in any one contract. Probably the leading concerns would be whether or not the design squad took adequate pains to work out the roadside items and whether the contractor/subcontractors have adequate knowledge and experience to know what they are doing.

### Utilities: Protection and Relocation

Two duties of the inspection team are to assure that the utility companies (or municipal departments) have been given adequate notice by the contractor prior to any work that might affect their installations, and that changes in locations of utility lines and appurtenances have been adequately charted, recorded, and provided to the utility owner. The utility owner is usually responsible for the adequacy of protection and for any approved work performed on the utility lines and appurtenances by the contractor.

## THE GRAY AREAS OF INSPECTION

If the inspector abides by the simple dictum of controlling quality without impeding progress, the results should be of a salutary nature. Reading, comprehending, and interpreting the contract documents (plans and specifications in particular), along with maintaining accurate records, are essentials to controlling quality. But nit-picking enforcement can be self-defeating when it impedes progress while, at the same time, accomplishes nothing of worth. If a low-value item has to be redone at high cost, a contractor may well attempt to retrieve the loss by cutting corners on some high-value items. Perhaps the one area where the inspector is justified in nit-picking is in the realm of safety, both to the workers and to the traveling public.

Another gossamer undescribed in most spec books is what to do when there are too many activities going on that cannot be adequately inspected by the available personnel. A rule of thumb states that inspection at the point of application is more important than inspection at the source or inspection in transit.

### Waste Disposal Sites

The inspector cannot look the other way when a contractor defiles an off-project site with waste products from the work being performed under contract. Although the contractor is required to dispose of wastes (hazardous and otherwise) in a proper manner and has the responsibility to do so, the agency can be held accountable for the contractor's ill-mannered actions off-site.

**Gratuities**

When a contractor offers small favors to inspectors, the practice is frowned upon. But there has to be a realistic rather than a sanctimoniously idealistic approach to the issue.

Let's begin with money—cash. In no way should the exchange of currency from contractor (or contractor's representative) to inspector take place. Well then, how about the traditional bottle of bourbon at Christmastime? Probably all right. The background for this type of gratuitous receipt on the part of the inspector traces back to statements made by former Illinois Senator Paul Douglas. Senator Douglas was held to have as high ethical standards as any political figure. When asked about small gifts and whether he would accept them, he stated that he thought it was all right if the gift did not exceed a value of five dollars. That was circa 1950. Five dollars does not go quite as far nowadays, but the scale of a small gift has been somewhat defined.

Another facet of this dilemma is the difficulty an inspector may encounter when refusing or turning down the offer of a small gift. The contractor may, unjustifiably, feel that the inspector has become something of an adversary and may not continue to work in a cooperative spirit. Many contractors believe that the small gratuity is nothing more than an expression of appreciation for an inspector's good judgment and not a payoff for past or future favors. Should an inspector still feel uncomfortable about receiving a small gift and equally uncomfortable about not accepting it, the following scene, adapted from real life, may offer some direction:

*Contractor:* "Hey, I've got a Thanksgiving turkey for you out in my pickup."

*Inspector:* "Good. Bring it in. I'd like to take it down to Father Forrester at the rectory. The church is sponsoring a kitchen for the homeless. They should appreciate a turkey dinner. My spouse has already bought our turkey, but thanks for making me look like a hero."

*Contractor:* "I'll go get the turkey right away."

In this case the inspector has both accepted the gift and not accepted the gift. It's likely that the contractor will understand.

**References**

Belangie, M. C. 1990. Factors affecting joint system performance. *Transportation Research Record 1268*:17–24.

NCHRP Report 284, *Evaluation of Procedures Used to Measure Cement and Water*

*Content in Fresh Concrete*. 1986. Washington, D.C.: Transportation Research Board, National Research Council.

NCHRP Synthesis of Highway Practice 51, *Construction Contract Staffing*. 1978. Washington, D.C.: Transportation Research Board, National Research Council.

Newman, Robert B. 1989. NCHRP Syntheses 145 and 146, *Staffing Considerations in Construction Engineering Management*, and *Use of Consultants for Construction Engineering and Inspection*. Washington, D.C.: Transportation Research Board, National Research Council.

O'Brien, James J. 1974. *Construction Inspection Handbook*. New York: Van Nostrand Reinhold Company.

TRB, State of the Art Report 8, *Guide to Earthwork Construction*. 1990. Washington, D.C.: Transportation Research Board, National Research Council.

# Part 5

# Operation

# 21
## Traffic

Much of the information in this chapter rightfully belongs in the chapters dealing with design. However, appreciation of traffic matters is considerably enhanced when the highway architect has a chance to play in traffic for a while. Inasmuch as there is an abundance of detailed literature on all the topics discussed, each entry following will be treated very lightly. The reader is urged to consult the references at chapter end for a more complete understanding of designing for and managing traffic operations.

## ACCOMMODATING THE FLOW OF VEHICLES

Directing traffic is more than a frenetic police officer gesticulating to autos coming at him from all directions at an intersection in downtown Rome. Signs, pavement markings, traffic lights, and overall geometry are employed to assist vehicle operators in getting to their destinations (hopefully) safely and efficiently.

### "Happiness is Five Green Lights in a Row"

The bane of all drivers is the red light that requires coming to a full stop. Delay and frustration follow to the extent that driving on a heavily travelled urban street is far from pleasurable. As mentioned in Chapter 13, converting busy intersections to free-flowing interchanges can help to smooth the traffic flow, but it isn't practical to do so at every block.

Signal coordination has become the traditional manner in which traffic engineers attempt to cope with downtown traffic flow. With the advent of computers and applicable software, signal coordination has been applied with good success in many places. *Time base coordination* for a heavily traveled linear corridor is applicable such that green phases immediately precede traffic platoons. Grid street systems require a greater degree of sophistication in computer technology. Sometimes success is marginal because the software program is continually robbing Peter (north-south) to pay Paul (east-west), or maybe the other way around.

## Signing and Delineation

Maintenance and law enforcement efforts have a continuing need for reference points on completed highways. There was a time when an accident report might have read: "Location of accident—four hundred feet south of the big rock near the old sycamore tree..." Nowadays, almost all highways have mileposts and other reference markings. *NCHRP Synthesis 21* (pages 6-12) outlines a few of the more common reference methods for locating points on a highway once the survey stationing is no longer available.

Signing and delineation are usually considered part of a typical design contract. Although there are occasions when these items are placed in a separate contract, more often than not it's worthwhile for the architect to have at least an acquaintance with such specialities. The "bible" again is the *Manual On Uniform Traffic Control Devices* (MUTCD). The manual has been organized into sections on signs (regulatory, warning, guidance, and motorist services), pavement markings, signals, traffic control islands, construction and maintenance, school areas, grade crossings, and bicycle facilities. (Caution: Haphazard use of the MUTCD could spell trouble.)

Even though the manual is considered gospel on the topic, the courts do not always recognize its divinity. A suit was brought against the Michigan State Highway Department several years back to recover injuries sustained during a skidding accident. The highway agency pointed out that the sign warning of the icy conditions fully complied with the MUTCD. The judge, however, said that, since the sign was up all year round, its value as a warning device was eroded. Eight areas where MUTCD may still have apparent deficiencies are outlined in *TRR 1114* (pages 16-18).

Signing must be subject to placement and information that drivers normally expect. Brackett (1992) outlines some sign deficiencies, and *FHWA-TO-86-1*, beginning on page 7, describes many of the essentials of the "Expectancy Concept." Unfortunately, there are agencies that have their own sign types and logos. Since a substantial number of drivers have difficulty understanding the standard traffic control devices (TCD) presently in use, it is probable that "nonstandard" traffic control devices (which in most instances have not been evaluated for their effectiveness) further increase driver confusion. Incorporation (or non-incorporation) of "nonstandard" type TCDs and logos is subject to the policy of the sponsoring agency. The architect has little to contribute in the way of exciting innovations or aesthetic license.

Insights into sign placement, materials, and legibility are contained in *TRR 1149*, pages 22-31. An interesting discussion of sign lighting can be found on pages 79-93 of *TRR 1111*, which leads us to our next topic—lighting.

## Lighting

Highway lighting is a distinct specialty. Nevertheless, the architect may wish to apply some of the information contained in the *Handbook of Safety Design and Operating Practices* (pages 71-82) as a framework for lighting requirements on smaller projects. Subsequent to publication, energy conservation concerns have effected policies in many parts of the country towards reducing the amount of highway lighting. However, reducing the intensity of lighting on an interstate highway appears (not surprisingly) to reduce the ability of drivers to detect objects on the pavement, according to *TRR 1149*, pages 1-7.

Reflectivity of light at nighttime and during heavy rainstorms also governs many sign and pavement marking materials. (For example, beaded pavement stripes are highly visible on a clear night, but virtually invisible during a night rainstorm.)

Traffic signal design, placement, and operation are best left to the specialists in the field, although this aspect may sometimes be included in the design package. If so, traffic signals (including locations, types, standards, specifications, and estimates) may be handled in the same manner as the highway architect deals with structural designs (explained in chapter 12).

## Disregard on the Part of Motorists

What percentage of the motoring public actually abides by the posted speed limit? Aside from speeding, common forms of noncompliance with posted traffic regulations include:

- Not coming to a stop at an intersection that is controlled by full stop signs;
- Encroaching on a pedestrian walkway when pedestrians are crossing;
- Playing games with a train and gates at a railroad crossing;
- Turning illegally;
- Running the red light; and
- Parking in a restricted zone.

Being that these actions are not endemic to a small segment of the driving public, there must be a realization that "perceived reasonableness" rules the actions of most persons. If signs, markings, and other traffic control measures appear to be reasonable, they are likely to be observed. All too often, however, traffic controls are excessively restrictive and, as a result, receive a high degree of noncompliance.

260   Highways: An Architectural Approach

**Other Roadside Hardware**

Guardrails, median barriers, crash cushions, and sign and luminaire supports provide a combination of protective measures and hazards. Hardware can be either flexible or rigid. Each situation requires an independent judgment as to applicability of the appropriate standard. Large trucks, standard size passenger cars, small cars, and motorcycles each react differently when impacting roadside hardware. How they collide is also to be considered. Troxel (1991) analyzes sideway impacts of vehicles colliding with roadside obstacles.

## HUMAN FACTORS

Members of the highway community are coming around to the realization that many of the people who use the magnificent roadways we design and build are, in all actuality, human beings. As such, people tend to behave as human beings. Some members of the highway community have observed that, unlike standardized lengths of culvert pipe or mass-produced autos, each human being is a somewhat unique individual with unique individual behavior patterns. So then, we could ask ourselves the question: "Can standardized highway operations accommodate unique individual behavior patterns of human beings?"

Answer: (silence).

With more than 40,000 humans killed and some 2 million maimed each year in the United States as a result of actions (by other humans) on our highway system, the answer to the above question may have to be: "Not yet." The highway architect need not feel guilty about taking a little extra time to determine if there are any unusual human factors that could enter into the many variables in a particular highway operation. Some incidental "human" items:

- Pedestrian accidents seem to occupy a disproportionate percentage of highway-related fatalities. The very young and the very old seem to be most vulnerable. (Guidelines for installing pedestrian crosswalks can be found on pages 19 and 20 of *TRR 1141*.)
- Providing information to motorists—mainly directional and safety oriented—at present is worded and positioned more to how the tort lawyer can use it in court, rather than how the driver of the vehicle can use the information on the road.
- Work zones (as already stated) continue to be hazardous, even at those locations where exceptional precautions have been taken to provide protection for motorists and workers. Constantly changing work zone situations combined with bewildered driver unexpectations and divided attention of workers create too many personal tragedies.
- Traffic accident data collection tends to disregard human factors.
- As with pedestrian accidents, the young and old vehicle operators are dispropor-

**FIGURES 21-1a and b.** Not a good location to place a pedestrian crossing! Drivers from the cross street must watch for traffic on the one-way main artery; when they see a clear space in traffic they turn into the pedestrian crossing and, because their attention has been directed upstream, drivers have virtually no expectation of pedestrian traffic. The problem can be solved simply by moving the pedestrian crossing to the upstream side of the cross street. By so doing, neither pedestrian nor cross street motorist need conflict.

tionate in their representation in highway accident statistics. For the older drivers, eyesight may be a major problem. Many elderly persons suffer from night myopia, where the pupil opens wider as the light dims, resulting in a severe loss of focus. A normal eyesight in the daytime may degrade significantly at night. Also, reflections from the sun in rear windows of vans and pickup trucks, high-beam headlights, and glare from mirrored surfaces, such as wet pavement and sign faces, can suddenly blind a driver to signs and pavement markings.
- Evaluation of high-hazard sites often overlooks human factors that might (through accumulation) create a dangerous condition miles from the sites—a dangerous condition that may seem to have no readily discernable cause. (Absence of slow-moving-vehicle lanes could be one example.)

There are some efforts underway to address these issues. *Positive Guidance*, an attempt to enhance operational efficiency and safety, joins the engineering and human factor disciplines to produce information systems matched to the characteristics of both driver attributes and high-hazard sites. *Positive Guidance* provides spot and short-segment sign, marking, and delineation improvements to increase safety and reduce operational problems based on the premise that a driver can often be given sufficient information to avoid accidents at hazardous locations. *FHWA Report TO-86-1* provides definitions and details of *Positive Guidance* and outlines some factors to consider along with a handy checklist (pages 29–32).

The highway community is not in total agreement with the FHWA definition of *Positive Guidance*, which tends to be site-specific. A total system definition is favored by the Council of State Governments and many insurance companies. Both definitions can be accepted by the highway architect without having to resort to doublethink.

## NONMOTORIZED TRAFFIC

Horse-drawn carts, canestoga wagons pulled by oxen teams and similar traffic of over a century ago will not be discussed here, but there are two recognized forms of nonmotorized traffic that are still with us and must be dutifully considered.

### Pedestrians

Although cited earlier as an aspect of noncompliance and also mentioned as a "human item," pedestrian crossings (marked as well as unmarked) are not taken seriously in many parts of the country: neither by motorists nor by pedestrians.

Physical impairments have to be factored into the agendums of pedestrian traffic. A person who uses a wheelchair for mobility isn't the only concern. A father pushing his child in a stroller; a mother guiding a toddler through an intersection; a Boy Scout aiding an elderly woman across the street; a young executive temporarily on crutches after a skiing accident: they all may find it difficult to cope with automobile traffic in the same manner as if they were not temporarily encumbered. Yet, to fully accommodate all pedestrian traffic as if everyone were disabled could restrict vehicle traffic to unacceptable levels. Compromises have to be reached. Best judgment again comes into focus.

### Bicyclists

Here we have a real problem. Many jurisdictions require that motorists allow the full lane width for a bicycle when there is no available bike lane or path. This is impractical when the bicyclist may be pedaling along at 15 MPH where the running speed of motorized vehicles is 45 MPH. To force the bicyclist over to the shoulder

**FIGURE 21-2.** To retrofit pedestrian overpasses to accommodate disabled persons, some agencies go to extensive measures.

or curb doesn't work too well either, because accumulated road dirt, stones, gravel, drainage grates, debris, the aerodynamic effects of passing traffic, and obtrusive roadside hardware make shoulders (or gutters) the most dangerous part of the roadway to cycle on.

Vehicle detectors in most jurisdictions are not designed to detect bicycles. In order to cross at an intersection controlled by an actuated signal, the cyclist must either violate the law or wait for a motorized vehicle to come along and actuate the green phase.

Operational aspects to provide for bicycle travel have yet to become widely accepted in the traffic engineering community.

## WEATHER CONDITIONS

A whole new menu is introduced when changes in the weather dictate modification of traffic operations. Fog, wet pavements, snow, and ice can deal havoc to the best efforts at providing safe and efficient transportation.

### Wet Pavements

Surface friction plays the critical role here. When a water film lubricates the surface, it reduces the direct contact between tires and pavement along with a concomitant reduction in available friction. Should the water film be supplemented by an oil film (common on many older paved surfaces), the available friction nearly disappears.

One may presume that the pavement surface that has the highest friction index when dry is also, comparatively, the least slippery when wet. But there are so many other factors (i.e., tire treads, superelevation, vehicle characteristics, operator reactions) that a conclusive determination of best operating practice is not possible. And should the pavement hover around freezing temperatures, the high friction pavement may become inordinately slippery because of the formation of ice particles in the asperities (surface roughness).

### Snow Removal and Ice

Although removal of snow and ice from pavements is usually performed by maintenance personnel and equipment, the function itself is one of operation. Unique problems are presented when the phenomenon of ice formation on bridge pavements occurs before icing takes place on the contiguous pavements approaching the bridges. Since the condition tends to be unpredictable, the responsible highway agency is often caught in the jaws of litigious sharks. If warning signs are placed at every bridge approach, they will soon be ignored. When, as cited in the Michigan case, they are up all year long, especially when the temperature is above 70 degrees farenheit, the vehicle operators can hardly be expected to take the warnings seriously. It's tough to handle the problem with simple signing and very expensive to treat the problem directly.

## MANAGEMENT OF TRAFFIC OPERATIONS

All sorts of schemes have been tried in attempts to provide both short- and long-term traffic management.

### Capacity

Increasing the capacity of a major arterial, expressway, or freeway during periods of high volume, lower level of service, and, consequently, lower running speeds of the vehicles can sometimes be attained by using the shoulder as an added lane. Effective accomplishment of such a tactic can open up the proverbial container of invertebrates. One approach is to allow only small cars or vehicles with passengers to drive on the shoulders. Enforcement problems may become extensive, and the access and egress to the regular traffic lanes can become restrictive at ramps.

As one may have already guessed, use of shoulders as part of the traveled way is common in California's major urban/suburban centers. Other states with heavy commuter congestion have treated this matter more gingerly. Yet there is some merit in its application if the negative factors can be overcome or, at least, assuaged.

*Safety and Operational Evaluation of Shoulders on Urban Freeways* (Urbanik, 1987), published by the Texas Transportation Institute, is a suitable guide for determining whether or not use of shoulders as part of the traveled way during certain periods of the day is worth considering.

**FIGURE 21-3.** Slow moving vehicle (climbing) lanes can do much to increase highway capacity in rolling country.

## Incident Management

An "incident" may be an accident, stalled vehicle, roadside diversion, or some other singular event that causes traffic to back up temporarily on a well-traveled route. Congestion resulting from an incident is quite different from that which occurs regularly on freeways and arterials that are incapable of handling the regular traffic demand. Congestion resulting from an incident, therefore, is a matter for operation; regular diurnal congestion is something to be handled by planners and designers.

Managing incidents requires more than just the highway agency. Police, fire departments, media, ambulance services, tow truck operators, and local government street departments may all be involved in any one incident. Table 21-1 outlines some incident management measures. Planning ahead for such contingencies requires a well-integrated management system. Here again, a highway architect can play the key role in getting an incident management system organized should one not already exist.

More on incident management can be found in FHWA SA-91-056; future courses of action will be addressed in Chapter 23.

## Hazardous Materials Shipments

When an incident involves a shipment of toxic wastes, the entire surrounding area may be at risk. For this reason, trucking of most hazardous materials is confined to specific routes (and, in some jurisdictions, specific times). Geographic information systems are now being used to locate routes that are practical yet removed from concentrated population centers.

TABLE 21-1  Measures for an Ideal Incident Management System (From NCHRP Synthesis 156)

| Need | Measures to Address the Need |
|---|---|
| Detecting and determining the nature of incidents. | Organize existing information sources into a comprehensive network for detection of incidents. |
|  | Design, build, and maintain an electronic surveillance and detection system. |
|  | Place closed-circuit television cameras along critical freeway links and at particularly troublesome locations. |
|  | Use other systems and procedures to gather all available information regarding what is happening on the freeway system. |
| A focal point for processing data and for collecting and disseminating information. | Establish a traffic operations center, appropriately equipped and staffed. |
|  | Use electronic displays, maps, or other means to visually depict freeway operating conditions. |
|  | Develop communications systems to receive and dispense information. |
| Active management of major incidents to speed removal of incidents and to manage traffic, to minimize congestion throughout the duration of incidents. | Establish procedures and working relationships to bring about coordinated response efforts by various agencies. |
|  | Form incident response teams. |
|  | Use truck-mounted variable message signs and highway advisory radio systems. |
|  | Design planned alternative routes. |
| Quick removal of incidents from traffic lanes. | Make use of service patrols that operate with vehicles capable of removing relatively lightweight vehicles from the freeway. |
|  | Have heavy service patrol vehicles and/or tow trucks available that are equipped to remove stalled heavy vehicles from traffic lanes. |
| Quick removal of major incidents. | Establish tow truck services to provide needed services in a timely manner. |
| Provide traffic information to motorists. | Use variable message signs located at key locations throughout the freeway system. |
|  | Use portable variable message signs that can be positioned and operated in conjunction with incident management. |
|  | Establish a highway advisory radio system, either ground-mounted or portable. |

| Need | Measures to Address the Need |
|---|---|
| | Develop a network of commerical radio stations to broadcast information and develop the means to quickly provide information to those radio stations. |
| | Create systems to provide information about long-term traffic conditions to print media. |
| Traffic management for construction, maintenance, and special events. | Institute the procedures and recruit the staff to develop traffic management plans for major activities. |
| | Organize an extensive public information effort for each major event. |
| | Apply incident management measures throughout the duration of each event. |

## Oversize Loads

Permits for oversize loads may be inadvertantly issued without the appropriate operational personnel (highway patrol, agency maintenance people, and safety officers) being informed. When this occurs, it's just another *mal de tête*.

## Energy Contingency Plans

Some local agencies have established strategies to deal with sudden petroleum shortages. The highway operations specialist needs to be aware of such strategies should they be implemented on short notice.

# RAILROAD CROSSINGS

Since a 45-car freight train barreling along at 60 MPH is unable to come to a rapid halt, its visibility is the key ingredient at a highway-railroad grade crossing. Sight distance from the highway must be adequate. Yet there are certain drivers who enjoy a challenge. They will elect to race the train to the crossing. When the race ends in a tie, the result is considered unacceptable.

Local practice coupled with agency and railroad policies proffer different warning systems for different traffic patterns—both vehicular and railroad. Because there are still inordinately high accident occurrences at many railroad crossings, it is evident that much improvement in operational practices is warranted.

## References

AASHTO. *An Informational Guide for Roadway Lighting.* 1976. Washington, D.C.: American Association of State Highway and Transportation Officials.

AASHTO. *A Policy on Geometric Design of Highways and Streets.* 1990. Washington, D.C.: American Association of State Highway and Transportation Officials.

AASHTO. *Guide for Selecting, Locating and Designing Traffic Barriers.* 1977. Washington, D.C.: American Association of State Highway and Transportation Officials.

*Arizona Bicycle Facilities Planning & Design Guidelines.* 1988. Phoenix: Facilities Planning Committee, Arizona Bicycle Task Force.

Botma, Hein, and Hans Papendrecht. 1991. *Traffic Operation of Bicycle Traffic.* Paper read at 70th Annual Meeting of the Transportation Research Board, 13-17 January 1991, Washington, D.C.

Bowman, B. L., J. J. Fruin, and C. V. Zegeer. 1989. *Handbook on Planning, Design, and Maintenance of Pedestrian Facilities, FHWA-IP-88-019.* McLean, Virginia: Federal Highway Administration.

Brackett, Quinn, et al. 1992. A Study of Urban Guide Sign Deficiencies. Paper read at 71st Annual Meeting of the Transportation Research Board, January 12-16, 1992, Washington, D.C.

Burnham, Archie C., Jr. 1990. *NCHRP Synthesis 162, Signing Policies, Procedures, Practices, and Fees for Logo and Tourist-oriented Directional Signing.* Washington, D.C.: Transportation Research Board, National Research Council.

Clarke, Andy. 1990. Gridlock 2020. *TR News,* January-February 1990:12-13.

FHWA. *Driver Expectancy in Highway Design and Operations, FHWA-TO-86-1.* 1986. Washington, D.C.: Federal Highway Administration.

FHWA. *Handbook of Highway Safety Design and Operating Practices.* 1978. Washington, D.C.: Federal Highway Administration.

FHWA. *Handbook on Freeway Incident Management, FHWA-SA-91-056.* 1991. Washington, D.C.: Federal Highway Administration.

FHWA. *Manual on Uniform Traffic Control Devices.* Regularly updated. Washington, D.C.: Federal Highway Administration.

FHWA. *Planning Design and Maintenance of Pedestrian Facilities.* 1989. McLean, Virginia: Federal Highway Administration.

FHWA. *Safety Cost-Effectiveness of Incremental Changes in Cross Sectional Design— Informational Guide.* 1987. Washington, D.C.: Federal Highway Administration.

FHWA. *Traffic Conflict Techniques for Safety and Operations.* 1989. McLean, Virginia: Federal Highway Administration.

Hange, William A., Jr. 1991. A central traffic control computer can save money. *ITE Journal* 61 (2):37-40.

*Highway Capacity Manual, Special Report 209.* 1985. Washington, D.C.: Transportation Research Board, National Research Council.

Institute of Transportation Engineers. 1982. *Transportation and Traffic Engineering Handbook.* Englewood Cliffs, N.J.: Prentice-Hall, Inc.

Mannering, Fred L., and Walter P. Kilareski. 1990. *Principles of Highway Engineering and Traffic Analysis.* New York: John Wiley & Sons.

NCHRP Legal Research Digest 6. 1989. *Impact of the Discretionary Function Exception on Tort Liability of State Highway Departments.* Washington, D.C.: Transportation Research Board, National Research Council.

NCHRP Report 330. 1990. *Effective Utilization of Street Width on Urban Arterials.* Washington, D.C.: Transportation Research Board, National Research Council.

NCHRP Synthesis 21. 1974. *Highway Location Reference Methods*. Washington, D.C.: Transportation Research Board, National Research Council.
NCHRP Synthesis 128. 1986. *Methods for Identifying Hazardous Highway Elements*. Washington, D.C.: Transportation Research Board, National Research Council.
NCHRP Synthesis 132. 1987. *System-wide Safety Improvements: An Approach to Safety Consistency*. Washington, D.C.: Transportation Research Board, National Research Council.
NCHRP Synthesis 156. 1990. *Freeway Incident Management*. Washington, D.C.: Transportation Research Board, National Research Council.
Pietrucha, Martin T., et al. 1990. Motorist compliance with standard traffic control devices. *Public Roads* 53 (4):131–38.
Robinette, Gary O. 1985. *Barrier-free Exterior Design*. New York: Van Nostrand Reinhold Company.
Ross, H.E. Jr., et al. 1989. *NCHRP Report 318, Roadside Safety Design for Small Vehicles*. Washington, D.C.: Transportation Research Board, National Research Council.
Dahir, Sabir H.M., and Wade L. Gramling. 1990. *NCHRP Synthesis 158, Wet-pavement Safety Programs*. Washington, D.C.: Transportation Research Board, National Research Council.
Transportation Research Circular 362. 1990. *Use of Benefit-Cost Analysis to Develop Roadside Safety Policies and Guidelines*. Washington, D.C.: Transportation Research Board, National Research Council.
Transportation Research Record 1111. 1987. *Traffic Accident Analysis, Visibility Factors, and Motorist Information Needs*. Washington, D.C.: Transportation Research Board, National Research Council.
Transportation Research Record 1114. 1987. *Traffic Control Devices and Rail-highway Grade Crossings*. Washington, D.C.: Transportation Research Board, National Research Council.
Transportation Research Record 1127. 1987. *Innovation, Winter Maintenance, and Roadside Management*. Washington, D.C.: Transportation Research Board, National Research Council.
Transportation Research Record 1133. 1987. *Roadside Safety Features*. Washington, D.C.: Transportation Research Board, National Research Council.
Transportation Research Record 1141. 1987. *Pedestrian and Bicycle Planning with Safety Considerations*. Washington, D.C.: Transportation Research Board, National Research Council.
Transportation Research Record 1149. 1987. *Visibility for Highway Guidance and Hazard Detection*. Washington, D.C.: Transportation Board, National Research Council.
Transportation Research Record 1160. 1988. *Traffic Control Devices*. Washington, D.C.: Transportation Research Board, National Research Council.
Transportation Research Record 1213. 1989. *Human Performance and Highway Visibility*. Washington, D.C.: Transportation Research Board, National Research Council.
Transportation Research Record 1225. 1989. *Highway Capacity, Flow Measurement, and Theory*. Washington, D.C.: Transportation Research Board, National Research Council.
Transportation Research Record 1233. 1989. *Design and Testing of Roadside Safety Devices*. Washington, D.C.: Transportation Research Board, National Research Council.

Transportation Research Record 1244. 1989. *Traffic and Grade Crossing Control Devices.* Washington, D.C.: Transportation Research Board, National Research Council.

Transportation Research Record 1247. 1989. *Visibility Criteria for Signs, Signals, and Roadway Lighting.* Washington, D.C.: Transportation Research Board, National Research Council.

Transportation Research Record 1254. 1990. *Traffic Control Devices for Highways, Work Zones, and Railroad Grade Crossings.* Washington, D.C.: Transportation Research Board, National Research Council.

Transportation Research Record 1264. 1990. *Transportation of Hazardous Materials.* Washington, D.C.: Transportation Research Board, National Research Council.

Transportation Research Record 1281. 1990. *Human Factors and Safety Research Related to Highway Design and Operations.* Washington, D.C.: Transportation Research Board, National Research Council.

TRB Special Report 218. 1988. *Transportation In An Aging Society.* Washington, D.C.: Transportation Research Board, National Research Council.

TRB State of the Art Report 6. 1987. *Relationship Between Safety and Key Highway Features.* Washington, D.C.: Transportation Research Board, National Research Council.

Troxel, Lori A., Malcolm H. Ray, and John F. Carney III. 1991. Side impact collisions with roadside obstacles. *Transportation Research Record 1302*:32-42.

Urbanik, Thomas, and C. R. Bonilla. 1987. *Safety and Operational Evaluation of Shoulders on Urban Freeways.* College Station: Texas Transportation Institute.

Zegeer, Charles V. 1988. *NCHRP Synthesis 139, Pedestrians and Traffic-control Measures.* Washington, D.C.: Transportation Research Board, National Research Council.

# 22

# Maintenance

There they go, toting picks and shovels, driving front end loaders and pickup trucks and graders, a so-so looking bunch in dirty clothes: the road crew. Sure, they appear something less than the crisp and polish seen on the parade grounds at West Point, but they are the most important people in the highway community.

Maintenance of the road system is labor-intensive. But don't judge those working folks—cement dust in their hair and bituminous splashings on their shoes—as lower caste. Particularly in the rural regions, these road crews are often comprised of community leaders and civic luminaries. It isn't unusual, on a country road crew, to find perhaps a school board member, maybe the mayor of a nearby village, a part-time newspaper reporter, the chairperson of the township garden and beautification council, and other highly regarded members of the community. One should not be too surprised if one of the equipment operators on the crew once served as county commissioner and is now a Green Peace activist. A job on the road crew is one of the more sought after, one of the most secure, and one of the better jobs out in the boondocks. With it goes the responsibility of making sure that the highway network is kept up to snuff.

Because of their high local profile, they take pains to go the extra effort, to be conscientious and dutiful. Neighbors and friends, being aware that they are on the road crew, can be quick to point out to them any deficiencies in the highway maintenance. Although the same characteristics do not always pertain to street crews in urban areas, there are many cases where even the city street fixers return splendid efforts to the taxpayers.

With this said, the highway architect will hopefully use this valuable resource— conscientious maintenance personnel—as input to many of the decisions that go into planning, design, construction, and operation. Visits to maintenance sheds, to crews at work on the road, and to district maintenance supervisor offices can give the architect many excellent suggestions on how to be an effective shepherd for any highway project.

## MAINTENANCE PRIORITIES

Some dozen or so activities account for more than 80 percent of a highway agency maintenance budget. In urban communities the efforts are directed mostly to bridge upkeep, pavement patching, curb and gutter sweeping, signs, signals, pavement markings, drainage functions, litter pickup, and, where climate requires, snow removal. Rural roadways receive the most attention in pavement repair, culvert cleanouts, snow plowing in cold weather, and mowing the roadside vegetation through the warm months.

### Environment

While performing routine activities, maintenance personnel are also cognizant that there are additional demands to comply with: new health, safety, and environmental regulations. To reach a long-term solution to compliance, highway agencies are continually working with health, safety, and environmental agencies at local, state, and federal levels, to identify costs of corrective measures, to improve maintenance management, to seek means for increased automation, and to be on the lookout for advanced technologies. It can be confidently stated that health, safety, and environmental considerations are high priority. There are also specific maintenance functions that take top priority.

### Patching

Because of incessant pressure from the public and media, pothole (chuckhole) repair is the first order of business with any road maintenance crew (see Fig. 22-1). Another driving force behind getting potholes patched before sundown comes from the legal fellowship. As part of a review of state highway agency liabilities, *Legal Research Digest Number 10*, from the National Cooperative Highway Research Program, asserts:

> The State is at all times under a duty to exercise reasonable care in respect to the maintenance and upkeep of its highways, and deriving therefrom is the obligation to inspect for and make timely repair of potholes and like defects appearing in the highway surface....It seems obvious that where State highway departments are faced with the problem of repairing many potholes (as following a severe winter) that not all the potholes can be repaired at the same time, and that the problem of which potholes

Maintenance 273

**FIGURE 22-1.** Repairing potholes is the immediate priority in maintaining roads.

to fix first, and which to repair later, must be made the subject of judgment or choice, and in accordance with the establishment of a schedule of priorities.

Wintertime patching is the biggest bugaboo. Rarely do cold weather patches, no matter what materials and methods used, hold up very long. When the weather warms up, the crew must go out and patch the same holes over (and over) again. It may not be the patches that are the problem; rather, it might well be the weakened structure around the patch that gives way. The roadway surface has become too weak through constant wear from pummeling by thousands of vehicles passing over it, or, as is often the case, the subgrade retains excessive moisture.

Although asphaltic pavements are more often in need of patching, the malady does crop up in older PCC pavements as well.

## Postponing Maintenance

The principal reason for pavements aging too rapidly is that (as mentioned near the beginning of this treatise) money budgeted for preventive maintenance or regular upkeep is skimpy—and occasionally it disappears. All too often, whenever there is a budget shortfall, the highway agency opts for the easy way out: defer maintenance.

Such a situation, where it occurs, must be recognized by the highway architect so that necessary safeguards can be incorporated into the design and the construction specifications. Should there be a local tradition of postponing maintenance after a highway is built, during the design phase the architect ought to explore use of materials, techniques, and methods to counter the eventuality of limited upkeep. Some inelaborate examples: specifying PCC pavements rather than asphalt; designing simple, unelegant structures; using straightforward (sometimes overdesigned) drainage schemes with a minimal number of bends; installing box culverts instead of CMP arches; and providing spartan rest area facilities. Most of such items will raise the cost of construction. Enough money somehow becomes available not to maintain but rather to build a more expensive project; or the sponsoring agency can get the extra money by snatching it from the maintenance budget. (But it never seems to work the other way around.) Obtaining the wherewithall to spend on highways does not always follow a logical path. Some analysts of highway economics have observed that assessment of road priorities in Europe and the Orient are driven by engineering principles, while in the United States the assessment is driven by financial sleight of hand. Reinforcing this observation is a statement made by Franklin E. White, Commissioner of the New York State Department of Transportation, in a 1990 statement to the United States Congress:

> ...under present guidelines, if we maintain bridges adequately, we have to do it with our own money. If we let them fall into disrepair so that they require major rehabilitation, the federal government helps pay for that rehabilitation.

The highway architect must learn to live with such monetary mischief.

## PAVEMENT ACTIVITIES

Considering that pavements are the core elements of highways, their maintenance commands center stage. Of the two principal classes of pavements—flexible and rigid (i.e., bituminous and portland cement concrete)—flexible bituminous concrete (asphalt concrete) requires the greater attention of highway maintenance forces.

### Bituminous Surfaces

Surface treatment and repair to asphalt pavements may take various and sundry modi operandi. From minimal-to-major effort, the following are the more common treatments and repairs:

- Pothole (chuckhole) patching;
- Crack sealing;
- Fog sealing;
- Rejuvenation;
- Slurry seal;
- Lane leveling;
- Chip seal;

- Sandwich seal;
- Micro-surfacing;
- Cape seal;
- Planing;
- Scarifying;
- Plant mix seal; and
- Resurfacing.

A quick study of how, when, why, and where for each of these procedures can help guide the architect while observing and discussing the applicability of each.

*Pothole (chuckhole) patching* must certainly go back some two millenia when Greek, Egyptian, and German slaves (properly supervised by their Roman masters) were assigned the task of keeping the Appian Way in first-class condition. Today's procedure is quite simple (assuming there are no problems with the subgrade): Clean out the hole; cut out the loose pieces of pavement around the hole; paint a tack on the edges of the pavement surrounding the patch; shovel in some patching material; level it off (slightly higher than the adjacent surface) with a rake; and compact it with a roller, truck tire, or tamper. If the hole is particularly deep, stone or gravel is often placed on the bottom before finishing off the top with patching material.

On roads that are far removed from the central cities, this operation might be performed by one or two persons with a pickup truck (a small pile of patching material and a can of tacking in the bed) and hand tools. In urban and suburban settings, repairs to a pock-marked street might be undertaken by a train of equipment (trucks, grader, roller, flashers) along with a crew of a dozen or more. A good number of the crew spend their time directing traffic.

*Crack sealing* is best done in cooler weather, when the cracks are open. The cracks are cleaned or routed to some extent by brooming or by pneumatic (air) hoses and then filled with a cutback, emulsion, or proprietary crack sealer to prevent moisture from penetrating to and into the pavement base. When completed, the black sealer contrasts with the lighter-colored asphalt concrete to give the appearance of a web spun by a drunken spider. However, if properly done, a crack-sealed pavement usually rides pretty well in spite of its raunchy appearance.

*Fog sealing* is the least expensive of the mechanized sealing practices and can be considered as preventive maintenance. After the pavement is broomed clean, a spread of oil is released from the spraybar on the back of a tanker truck as the tanker passes along the pavement. Since there is no aggregate involved, the application rate must be low to avoid tackiness and loss of skid resistance. Fog seals are only effective when started early in the pavement life (1-3 years); when the pavement shows no signs of distress, and when there is but minor loss of surface fine aggregate.

*Rejuvenation* is much like fog sealing. Its purpose is to extend the time period between scheduled seal coats. The asphaltic material used is high in the maltene fraction, application rates are slightly higher than fog sealing, and sand is spread over the oil in order to retain a modicum of skid resistance. Neither fog seals nor rejuvenation will fill anything wider than hairline cracks; therefore, more extensive treatment is necessary for pavement surfaces that are beginning to show signs of distress.

Next up in the hierarchy are *slurry seals*, a blend of emulsified oil, small aggregates, and (sometimes) special additives. Proportioning and mixing these components requires special equipment and immediate placing, followed by a 2-to-12-hour curing period during which traffic may not tread upon it. Slurry seals give a nice appearance and can be used on pavement surfaces that are beginning to show hints of aging. It is a good practice to patch any potholes and fill the wider cracks before attempting a slurry seal. Some caveats must be observed when slurry seals are applied. Researcher Elton Ray Brown, doing a study for the U.S. Army Corps of Engineers and the U.S. Air Force, described the following:

> Slurry seals are normally used on pavements subjected to low traffic volumes. When slurry seals are subjected to high traffic volumes, the life is greatly reduced. One of the biggest problems observed with slurry seals is the loss of bond between the slurry and the underlying asphalt mixture. This loss in bond can be a result of several factors which may occur simultaneously or separately. The underlying surface must be clean prior to application of a tack coat. Construction equipment operating on a pavement to be sealed can often track mud and other foreign material onto the pavement surface preventing the development of a satisfactory bond. The tack coat is normally applied immediately prior to applying the slurry so dust or other debris on top of the tack is not a problem. Many times the slurry is placed at a time when the temperature is relatively low resulting in poor bond. Slurry seal should be performed in hot weather. It generally should not be placed when the ambient temperature is below 50 or 60 degrees Fahrenheit. For best results, slurry seals should be rolled with a rubber tire roller after placement....

Professor Brown goes on to state:

> Performance data determined from a number of Air Force Bases has shown that slurry seals typically last for 3-6 years depending on the construction quality and environmental conditions.

On highways, slurry seals are not expected to have this long a life-span.

When the maintenance forces are required to attend to ruts, one of the more useful measures is *lane leveling*. In its simplest form, the rutted areas are tacked, asphaltic concrete (or a suitable proprietary product) is placed in the ruts, and a long-wheelbase grader (preferably with a skilled operator) screeds the material level with or slightly higher than the unrutted surface. Rolling or other satisfactory compacting completes the operation. Usually, a seal coat is placed soon after lane leveling.

A most cost-effective asphalt pavement maintenance treatment is the *chip seal*.

The procedure begins in a similar manner to fog sealing except that the asphaltic oil is gooier and applied more heavily (occasionally pre-primed with a lower viscous oil). Immediately after the oil goes down, the chips consisting of $1/4$-inch or $3/8$-inch or $1/2$-inch stones (all of the same size) are spread evenly and rolled into the emulsion.

The biggest complaint lodged against chip seals is by fast-driving motorists who report smashed windshields from flying stones during the first few weeks after a chip seal. Even the most concentrated and complete rolling of the stones rarely prevents this problem. Although steel wheel rollers are occasionally employed, their use is discouraged, since they tend to break up some of the aggregate and they bridge over transverse low spots; pneumatic rollers are preferred.

Many highway agencies prescribe, on two-lane rural roads, a program to chip seal every 5-6 years whether the surface needs it or not. Such a schedule is generally regarded as optimum preventive maintenance. Some agencies extend the time period to 8-9 years by including two intermediate rejuvenations between seal coats. Spot sealing just a short section of deteriorating roadway (500 feet or so) before the section starts to go can also be effective preventive maintenance.

If clean aggregates are available, a *sandwich seal* is also cost-effective. This procedure is the same as a chip seal except that, after tacking, a course of slightly larger aggregate is placed prior to applying the asphaltic cement and surface chips. In effect, it is a double chip seal that lasts nearly twice as long with just one pass of the tanker (not including the tacking) and half the cost of the asphaltic cement.

*Micro-surfacing* employs polymers added to a slurry seal mix. By so doing, the thickness of the slurry can be significantly increased. Filling of ruts and cracks can be done in one swell foop, but the price is high because of special equipment needed and higher material costs.

Topping off a chip seal with a slurry seal has come to be known as a *cape seal*. Advantages include a long service life and eliminating some of the cracked windshields resulting from flying chips. Disadvantages are the longer time involved to complete the work, keeping traffic out of the work area, and the relatively high cost.

When a surface begins to undulate so that the riding qualities emulate those of a 22-foot yacht crossing the North Atlantic, *planing* the surface could offer a short-term fix. Use of a motor grader or, better still, a heater-planer machine may be able to restore the pavement level to what it was before pushing, shoving, corrugations, and rutting created the undulations. Patched areas over utility trenches (which may have used materials different from that in the adjacent roadway) are particularly susceptible to pushing, shoving, and corrugating. They may have to be completely removed and replaced as part of the operation.

Going one step further, by adding a scarifier to a heater-planer, the top layer of the pavement can be recycled in place to provide a longer lasting surface. *Scarifying* is best used where bleeding and flushing have occurred as a result of too much asphalt or instability in the mix.

*Plant mix seals* are essentially minor resurfacing efforts. They are rarely performed by regular maintenance crews, more often by contract construction. But the money comes out of the maintenance budget.

*Resurfacing* of a well-traveled asphalt road becomes a major operation and can be classified as a betterment rather than a routine maintenance activity. There are times when there is no alternative to resurfacing; other times it becomes necessary because regular upkeep was not properly tended to.

Preventive maintenance, through one or more of the procedures described above, is required to counter the hardness, brittleness, and loss of adhesive power of the asphaltic cement as it wears and ages.

Reconstruction—including correction of drainage problems, restoring sections that have developed a weak subgrade, placing a new base course, and resurfacing—is often the only way to go when a length of roadway shows more than moderate levels of distress (i.e., alligator cracks, wheelpath ruts, poor ride index, significant longitudinal cracking and/or map cracking and/or transverse cracking). This activity is generally not considered maintenance, but becomes mandated when deferred maintenance receives top priority.

## PCC Repairs

Portland cement concrete pavements are a different maintenance concern than asphalt pavements. For the first few years they require hardly any maintenance effort at all (not counting snow removal). Then, all of a sudden, they might explode in hot weather, joints crack open, spalling takes place, and they seem to age in a hurry. How about preventive maintenance?

Joint and crack sealing is done in a manner similar to that for bituminous pavements, using different sealing materials. Just about all of the other preventive maintenance steps that can be taken fall into the categories of rehabilitation or reconstruction. They may include rotomilling to even up the surface at the expense of reducing the slab thickness, bonding PCC overlays (thick or thin), mud-jacking settled slabs, cracking-and-seating followed by placing a course of asphalt concrete over the top, or just removing the spalled and broken slabs and replacing them in kind.

## Shoulders and Sidewalks

Sometimes shoulders are treated as part of the pavement and sometimes not. It depends on the shoulder makeup. If the shoulders are but a striped extension of the asphalt or PCC, they are maintained as part of the traveled way. Conversely, shoulders of gravel, clay, grass, soil cement, or thin layer of asphalt may require individual attention. Although initial construction costs for weak shoulders are lower, maintenance gets to be expensive after a few years. The upkeep procedure is essentially to restore portions of damaged shoulders to their original condition. Where sidewalks and curbs are part of the maintenance responsibility, the principal concerns are directed towards reducing heaving by tree roots, removing low-hanging tree limbs, and eliminating puddles at crosswalks.

## Pavement Markings and Stripes

Most maintenance organizations use special crews to place stripes and special markings, such as pedestrian crossings and turning lane messages. These crews have been trained and have practiced the art so that the results are neat and tidy (see Fig. 22-2). In the northern climates, where plowing, sanding, and salting provide bare pavements throughout the winter, stripes and other pavement markings are pretty well obliterated by springtime. The paint crews must get the markings back on the pavement in rapid order. In jurisdictions where lines and markings are pasted on (e.g., preformed polymer pavement markings), the same problem must be addressed at the end of the cold season. Sunshine also has a debilitating affect on pavement markings; however, replacing the lines isn't as urgent in the southern states because of the more gradual fading.

## Gravel and Unsurfaced Roads

A motor patrol (grader) is the tool used most often to keep low-travel non-hard-surface roads passable. Skill and experience on the part of the operator are the key ingredients to success. The critical factor is to make sure that the water from storms and from spring runoff is diverted to ditches and drainage culverts and is not allowed to flow over the top of the road, except at designated dips in the road profile that correspond to crossings of natural flow lines.

**FIGURE 22-2.** Clear pavement markings are a visible sign of attention given to highway maintenance.

Maintenance personnel may be required to treat unsurfaced roads by spreading calcium chloride or special enzymes and mixing them into the top layer, or to upgrade the surface marginally by adding some oil, as in the three-oil treatment.

### Winter Maintenance

Spreading sand and one or more forms of salt, together with plowing the snow off to the side in rural areas or loading snow on trucks downtown, can chew up a significant portion of the maintenance budget in northern states. Not only must streets and roads be cleared, but access ought not to be impeded where there is measurable pedestrian activity.

Because of its noncorrosive and environmentally acceptable properties, calcium magnesium acetate is replacing road salt as a deicer in well-heeled municipalities; its cost is very high when compared to sodium chloride, but sometimes local environmental considerations require limited use and restricted storage of road salt. Other proprietary materials to replace road salt are also under development. If a highway architect enjoys thrills such as sky diving or the monster turn-your-stomach-inside-out rides at the amusement park, a comparable thrill can be enjoyed by riding shotgun with the regular snow plow driver clearing a narrow, winding mountain highway in the middle of a zero-visibility blizzard.

### Rock Patrol

In hilly country and canyon locales, a maintenance crew may keep a snow plow on one of their trucks all summer. This truck will be used to shove any rocks off the traveled way during a daily patrol of sections where rocks are known to fall onto the pavement.

### Clearing Slide Areas and Hazardous Spills

Sometimes, after a heavy rainfall or rapid snow melt, the side of a hill slides down and covers a portion of the road. Or maybe a tanker overturns on the freeway and begins oozing a funny-colored odd-smelling fluid. Out goes the crew to clean up the mess.

## DRAINAGE

Next in importance to keeping the traveled way serviceable, is the upkeep of the drainage system. When the maintenance shed foreman is provided with a set of as-constructed plans for each project in his or her jurisdiction, it makes it a lot easier for him or her and the crew to find ways to keep the water flowing downhill and to prevent washouts.

## Cleaning the System

Catch basins, drop inlets, pipes, culverts, gratings, channels, and ditches have to be inspected and freed of debris at periodic intervals. Several months after riding the snow plow in the blizzard, the thrill-seeking highway architect can gain yet more excitement by accompanying the drainage cleanup crew into the manholes and culverts and join them in clubbing Norway rats, dodging water moccasins and coral snakes and rattlers, slapping off fire ants, beating angry wolverines, swatting alligators, and similar nuisances while going about their duties. (There's no end of thrills for highway maintenance people.) Drainage maintenance in urban areas also includes cleaning gutters on a regular basis.

## Rehabilitation

Many times the road crew will be called upon to perform some rehab to the system. Since wear in pipes occurs primarily to the invert (bottom), paving the invert is a simple means to extend the service life of a pipe. Sliplining (sliding a slightly smaller pipe inside a worn pipe) is often done when a drainage scheme is overdesigned. More expensive and specialized rehabilitation, such as relining, shotcreting, inversion lining, and complete replacement, are usually handled on a contractual basis as construction rather than routine maintenance.

## Slope Protection and Landscaping

If a good landscaping scheme is properly maintained, there is small likelihood that there will be many slope problems. Since maintenance personnel realize this, they generally take good care of the plantings. There is one problem recurring in some jurisdictions and it's not usually the fault of the maintenance personnel, but rather those higher up in the pecking order of the highway agency, whose fear of the accountants leads them into rash acts. The problem is that, after installing a good landscaping scheme and getting the plants started with watering or irrigation, to save money, the water is turned off and the plants left to die of thirst during the first drought season. The highway architect must be aware of this possibility when specifying any slope plantings that require watering during sustained dry periods.

# BRIDGES AND MAJOR STRUCTURES

If it isn't the accountants, it's the lawyers. One can understand the extent of liability that a highway agency sustains when a large span collapses into the river below. But there are less spectacular instances where the charge is made that a structure or portion thereof constituted a hazard and litigation proceeds in the prescribed manner at the standard pace. Who in the highway agency gets the blame?

282    Highways: An Architectural Approach

Maintenance, of course. Sovereign immunity is reserved for the planners "to protect the decision-making process." The function of maintenance is not decision making, so the tort attorneys go after the issue loosely described as "inadequately maintained."

"Adequate maintenance" becomes an annoying task when bridge engineers do not take maintenance procedures and methods into consideration as they design structures. Just how to examine out-of-the-way tension members, to patch or remove rusty niches, or to paint some hidden structural components can become a real mystery when access to certain critical members is restricted. Thus, we find that bridge inspection has to precede the actual fixing up of potentially dangerous or crucial conditions (see Fig. 22-3).

**FIGURE 22-3.** Bridge inspection isn't always easy. Even an NBA center would have difficulty examining the underside of this structure without some form of mechanical aid.

### Inspection Techniques

Quite a few different approaches are used to check out a bridge. The more common include:

- Visual;
- Dye penetration;
- Acoustic crack detection;
- Magnetic crack defining;
- Ultrasonic testing;
- Magnetic particle testing;
- Acoustic emissions;
- Radiography; and
- Robotic video.

A few of the items that need to be regularly inspected and sometimes fixed include framing systems, joints, hanger connections and supports, pins, bearings, rocker arms, welds, rivets, exposed reinforcing bars, and footings subjected to hydraulic scour. Adequate redundancy must also be determined during inspection debriefing.

It's easy to see that bridge inspection and maintenance are not done by the regular road crew, except for a few mundane chores such as removing graffiti. Some highway agencies in urban locales have large and very specialized divisions to take care of their structures. Others farm the work out.

The FHWA *Bridge Inspector's Training Manual* is a good primer on the subject. And, not to forget, governmental regulations are something else to contend with. For example, proper containment (and disposal) of lead-based paints is yet another environmental consideration.

## ALONG THE ROADSIDE

Many of the miscellaneous activities performed by the road crew are sandwiched between those activities just described. They need not be done in a haphazard fashion, but will generally fall to a lower priority and, therefore, may not get all the attention that a first-rate maintenance operation requires.

### Mowing and Trimming

Roadside grass between the shoulders and right-of-way fence needs to be kept tidy not only for aesthetic reasons, but also to assure that adequate sight distance is continually provided. There are instances where poor mowing practices lead to introduction of weeds where the grass and legumes originally sowed along the right-of-way no longer persist (see Fig. 22-4). The highway architect needs to assure

**FIGURE 22-4.** Roadside mowing practice depends on seasonal phases as well as vegetation types.

that proper mowing practices (i.e., height of cut, use of appropriate equipment, months when grasses are to be fertilized and mowed) are followed by maintenance personnel.

Although it is the road crew who must trim bushes and remove tree branches to keep adequate clear zones, the determination of which branches and how much trimming usually falls to the traffic and safety experts. If, for instance, a particular tree on a highway segment is continually the object of errant vehicles, its removal may be ordered even if it falls outside the prescribed clear zone. The architect ought to be involved in describing the methods for such removals.

## Chemical Control

Controversy is at the core of what chemicals to use on noxious plants within the highway right-of-way. Not only the effect on the surrounding environment, but also the protection of those using the chemicals is at issue. It appears to many maintenance specialists that the rules for handling exotic chemicals keep changing at a faster pace than women's fashions.

## Rest Areas

Another factor that can eat into a maintenance budget is that of keeping rest areas in good order. Different jurisdictions use different approaches in their upkeep. In some places the work is farmed out to a custodial service; other places have a full- or part-time residency on site; still others use the highway agency crews supplemented by specialists (plumbers, electricians, and building tradesmen). Three of the top problems in keeping rest areas serviceable are vandalism, trash dumping, and littering. A ball park guess at whether a highway agency expends enough effort in maintaining rest areas, if on-site inspection is not feasible, is to determine how much money is spent per user. A North Dakota analysis found that the cost could vary from $0.06 to $1.58 per user. The average was $0.14 per user in 1984. Therefore, if a highway agency is only allocating $0.14 or less per user, the assumption can be made that rest area maintenance is not especially good. The architect should take this into account when designing any new or remodeled facilities.

## Litter Pickup

If one hasn't been involved in cleaning up the litter that accumulates hourly along the side of a highway, he or she may be unpleasantly surprised to learn the magnitude of the problem; the total tonnage collected by maintenance personnel can be staggering. Many community groups are now adopting nearby highway segments and providing some of the cleanup service on a voluntary basis. Nevertheless, proper disposal of roadside litter, as well as dead animals, usually becomes the responsibility of the regular road crew.

## Guardrail

Nicks, dents, and bends in guardrail, damage to attenuator barrels, and bruising of concrete barriers are usually repaired by the road crew either on-site or back at the shed (station).

## Lighting and Signals

Urban and suburban maintenance needs to give continual attention to keeping traffic signals and roadway lighting fully operational. Sometimes this work is taken over by the local utility company.

## Signs and Delineators

Broken, bent, graffitied, stolen, and otherwise damaged signs and road markers have to be replaced. Signs can't be removed and fixed back at the shed because their messages must be presented at all times.

## Fence Repair

Cattle, vandals, and errant autos may render right-of-way fences unserviceable, whereupon the maintenance crew is called upon to perform the necessary fixing up.

**FIGURE 22-5.** Access to gantry signs requires special equipment and attention to traffic control during maintenance.

## MAINTENANCE MANAGEMENT

There was a time when maintenance activities were conducted helter skelter. Beginning in the late 1960s, a number of state highway agencies (notably Iowa and Louisiana) began to manage their maintenance activities to gain greater efficiency and get better returns on their maintenance outlays. Now, virtually all highway agencies—even those at the local level—have adopted the principles of maintenance management:

- Developing work programs;
- Budgeting (with caveats cited above);
- Scheduling work;
- Reporting; and
- Evaluating performance and costs.

## IN CONCLUSION

If a little bit of knowledge is a dangerous thing, and a very little bit of knowledge a very dangerous thing, the reader is now extremely dangerous in the discipline of highway maintenance.

**References**

Barksdale, Richard D. 1991. *NCHRP Synthesis 171, Fabrics in Asphalt Overlays and Pavement Maintenance.* Washington, D.C.: Transportation Research Board, National Research Council.

Bowman, B. L., J. J. Fruin, and C. V. Zegeer. 1989. *Handbook on Planning, Design, and*

*Maintenance of Pedestrian Facilities*, FHWA-IP-88-019. McLean, Virginia: Federal Highway Administration.

Brown, E. R. 1988. Preventive Maintenance of Asphalt Concrete Pavements. Paper read at the 67th meeting of the Transportation Research Board, January 1988, in Washington, D.C.

Burns, E. Nels. 1990. *NCHRP Synthesis 170, Managing Urban Freeway Maintenance*. Washington, D.C.: Transportation Research Board, National Research Council.

Cunard, Richard A. 1990. *NCHRP Synthesis 157, Maintenance Management of Street and Highway Signs*. Washington, D.C.: Transportation Research Board, National Research Council.

FHWA. *Structural Materials Evaluation Laboratory, RD-90-073*. 1990. McLean, Virginia: Federal Highway Administration.

*Field Manual on Design and Construction of Seal Coats*. 1981. College Station: Texas Transportation Institute.

Flynn, Larry. 1991. Cape sealing technique thwarts loose rock woes. *Roads and Bridges* 29(12):28-32.

Harland, J. W. et al. 1986. *Inspection of Fracture Critical Bridge Members*, FHWA-IP-86-26. McLean, Virginia: Federal Highway Administration.

King, G. F. 1989. *NCHRP Report 324, Evaluation of Safety Roadside Rest Areas*. Washington, D.C.: Transportation Research Board, National Research Council.

Kitamura, Ryuichi, Huichun Zhao, and A. Reed Gibby. 1989. *Development of a Pavement Maintenance Cost Allocation Model*. Davis, California: Transportation Research Group, University of California at Davis.

Miller, C. R. 1989. *NCHRP Synthesis of Highway Practice 148, Indicators of Quality in Maintenance*. Washington, D.C.: Transportation Research Board, National Research Council.

Marti, M. M. and E. L. Skok, Jr. 1984. *Surface Condition Rating System, Field Guide*. St. Paul, Minnesota: Midwest Pavement Management, Inc.

NACE. 1986. *Action Guide, Maintenance Management*. Ottumwa, Iowa: National Association of County Engineers.

NCHRP Legal Research Digest 10. 1990. *Liability of the State for Injury-Producing Defects in Highway Surface*. Washington, D.C.: Transportation Research Board, National Research Council.

NCHRP Legal Research Digest 14. 1990. *Liability of State Highway Departments for Defects in Design, Construction, and Maintenance of Bridges*. Washington, D.C.: Transportation Research Board, National Research Council.

NCHRP Program Report 333. 1990. *Guidelines for Evaluating Corrosion Effects in Existing Steel Bridges*. Washington, D.C.: Transportation Research Board, National Research Council.

NCHRP Program Report 334. 1990. *Improvements in Data Acquisition Technology for Maintenance Management Systems*. Washington, D.C.: Transportation Research Board, National Research Council.

NCSPA. 1990. Rehabilitation can be cost effective. *Pipe Line*, Spring 1990. National Corrugated Steel Pipe Association.

O'Brien, Louis G. 1989. *NCHRP Synthesis 153, Evolution and Benefits of Preventive*

*Maintenance Strategies.* Washington, D.C.: Transportation Research Board, National Research Council.

*Redevelopment Plan for Highway Rest Areas in North Dakota.* 1984. Bismarck: North Dakota Department of Transportation.

Thomas, H. Randolph, and David A. Anderson. 1990. *Evaluation of Experimental Cold-Stockpiled Patching Materials for Repairs in Cold and Wet Weather.* Transportation Research Record 1268:52-58.

Transportation Research Record 1205. 1988. *Pavement Maintenance.* Washington, D.C.: Transportation Research Board, National Research Council.

Transportation Research Record 1246. 1989. *Winter Maintenance, Roadside Management, and Rating Routine Maintenance Activities.* Washington, D.C.: Transportation Research Board, National Research Council.

Transportation Research Record 1259. 1990. *Chip Seals, Friction Courses, and Asphalt Pavement Rutting.* Washington, D.C.: Transportation Research Board, National Research Council.

Transportation Research Record 1300. 1991. *Asphalt Pavement and Surface Treatments: Construction and Performance.* Washington, D.C.: Transportation Research Board, National Research Council.

U.S. Department of Transportation. 1979. FHWA *Bridge Inspector's Training Manual.* Washington, D.C.: Federal Highway Administration.

# 23

# Intelligent Vehicle/Highway Systems (IVHS)

Driving tasks are getting to be more mechanized as autos come equipped with sundry computerized gadgets. Cars, buses, and trucks are becoming more "intelligent"; they take care of chores that used to be part of the driver's duties. Intelligent vehicles require smart highways. Before the end of the twentieth century, Intelligent Vehicle/Highway Systems (IVHS) will be recognized members in the makeup of the landscape.

IVHS is not exactly new. At the 1939–40 New York World's Fair, the most popular of all the attractions was the General Motors exhibit, the centerpiece of which was a demonstration of IVHS called *Futurama*. Fair visitors were given a tour around a large scale-model of a futuristic urban center and its animated highway network as the recorded announcer described how advanced traffic management and automated vehicle control would be applied on future highways to enhance efficiency and safety of transport. Lane change maneuvers, ramp entrance and exit options, and speed of travel were all to be directed from overhead traffic control centers straddling each highway. It appeared that the automobile operator would have little to do, aside from reading the newspaper or eating lunch. One-half century later, such a concept is under development, in Europe and Japan, and under study in the United States.

## A CONCEPT COMING TO FRUITION

The 1930's vision comes in four generic systems:

- Advanced traffic management;
- Advanced driver information;

290   Highways: An Architectural Approach

- Automated vehicle control; and
- Freight and fleet operations.

All four systems must be interactive with each other for IVHS to be effective. Why adopt all these expensive concepts?

One easily understood reason was stated in the March 1990 Report to Congress on Intelligent Vehicle/Highway Systems by former Transportation Secretary Samuel K. Skinner:

> ...Many drivers do not have the reflexes to take appropriate action in emergency situations. Of particular note is the growing number of older drivers. These drivers have slower reaction times and poorer vision than younger drivers.

What it boils down to is survival on the highways for all of us.

### Advanced Traffic Management

Means for highway agencies to monitor traffic conditions, adjust traffic operations, and respond to incidents are becoming more sophisticated and technological. Already in place we find vehicle detectors embedded in the pavement, coordinated traffic signalization, changeable message signs, and, as shown in Figure 23-1,

**FIGURE 23-1.** Centralized TV monitoring of rush hour traffic in Minneapolis. [Photo: Minnesota Department of Transportation.]

centralized TV traffic monitors. Under study are some Big Brother initiatives: automated vehicle identification, road pricing, hazardous cargo restrictions, and alternative methods for law enforcement.

### Advanced Driver Information

What is the best routing to get to a destination? Advanced driver information will eventually consider congestion, accidents, weather conditions, construction site lane restrictions, road conditions, running speeds or speed limits, and other information feedbacks, to assist a vehicle operator in planning a trip. In addition to pre-trip electronic route planning, advanced driver information technology is also expected to embrace:

- On-board replication of maps and signing;
- Traffic information broadcasts tailored to more specific sites than the traditional commuter sky watch;
- Safety warnings (e.g., "You're getting too close to the vehicle in front of you");
- Navigation systems on-board where lane changes and similar maneuvers can be coordinated; and
- Route guidance while traveling.

### Automated Vehicle Control

Here is where HAL takes over and the "operator" no longer needs to drive. We already have the genesis of automated vehicle control with cruise control and antilock braking systems on many autos. Add to these such features as radar braking, automatic headway and lateral control, adaptive speed regulation, and an automated highway system in order to bring automated vehicle control on line.

### Freight and Fleet Operations

Compliance with regulations drives the implementation of strategies directed at heavy vehicle and commercial endeavors. Vehicle inspections, cargo contents, weight restrictions, routing options, and a host of other governmental concerns can be handled more expeditiously when advanced technology is applied. The private sector (i.e., delivery vehicles, trucking companies, and express operators) will doubtless benefit from the increased efficiency offered by IVHS.

### Acceptance of IVHS

Culturally, Germany and Japan have traditions that encourage acceptance of technologic advances that are generated by directive of the designated authorities.

In the United States, acceptance of anything promoted by government is traditionally resisted. How long it will take for the essential IVHS elements to be put in place may depend on cooperation between private enterprise initiative and governmental regulation. Highway architects will then have many more tools with which to ply their profession.

**References**

Allen, R. Wade, et al. 1991. Laboratory Assessment of Driver Route Diversion in Response to In-vehicle Navigation and Motorist Systems. Paper read at 70th Annual Meeting of the Transportation Research Board, January 13-17, 1991, Washington, D.C.

*Discussion Paper on Intelligent Vehicle-Highway Systems.* 1989. Washington, D.C.: U.S. Department of Transportation.

Pollard, William S., Jr. 1990. Determination of Least Cost Approach to Congestion Reduction in Urban Areas. Paper read at the Mountain Plains Consortium Transportation Research Conference, November 29, 1990, Salt Lake City, Utah.

Saxton, Lyle G., and G. Sadler Bridges. 1991. Intelligent vehicle-highway systems, a vision and a plan. *TR News* 152:2-6,32.

Spreitzer, William M. 1990. Technology, vehicles, highways and future transportation. *AASHTO Quarterly* 69(1):4-5,25-33.

Tarnoff, Philip J., and Ted Pugh. 1990. *NCHRP Synthesis 165, Transportation Telecommunications.* Washington, D.C.: Transportation Research Board, National Research Council.

# Index

AASHO (American Association of State Highway Officials), predecessor to AASHTO, 56, 75
AASHTO, *see* American Association of State Highway and Transportation Officials
Abson method, 251
Access, 29, 118, 166, 200-201. *See also* Right-of-way
   control, 7, 200
   limited, 16, 200
   provision of, 20, 235, 280
   restricted, 27, 264
Accidents, *see* Safety
ADT, *see* Average daily traffic
Advanced traffic management, 289-290
Advanced driver information, 289, 291
Aerial photography, 40, 50
Aesthetics, 69-70, 114, 140, 258
   alignment, 91-92, 103
   grading, 224, 244
   excavation, 126, 224
   fencing, 182
   slopes, 116, 134, 139, 224

   structure, 144, 154
   utility, 186
   vegetation, 176, 283
   visual, 97-100, 134, 144
Aggregates, 55, 57, 62, 78-84, 88-89, 227, 275-277
Alignment, 65, 76, 86, 91-92, 97, 103-121. *See also* Control line; Geometry
   affected by waterway, 65-68, 93-94
   coordination, 105-106
   correcting, 92
   horizontal, 97, 104-107, 109-113, 161-168
      spiral curve, 107, 113
   pipe, 246
   stream, 69-70
   vertical, 66, 97, 104-106, 108-113, 116-117, 161
Alternatives, 14-22, 38, 40-41, 44, 193
American Association of State Highway and Transportation Officials (AASHTO), xv, 99, 113, 208
Arch bridge, 147-148

Arterial route, 169, 195–196
Asphalt, see Bituminous materials
　batch plant, 100
　concrete, 57, 72, 78, 210, 276
　cutback, 60–61, 275
　emulsion, 60–61, 275, 277
　Institute, 61
　pavement, 76–80, 82, 88, 228–229
　　compacting, 244
　　financing, 99
　　inspecting, 249–251
　　joints, 228
　　resurfacing, 278
　　rutting, 79–80
　　testing, 249–251
Automated Vehicle Control, 290–291
Average daily traffic (ADT), 7
　average annual daily traffic (AADT), 7
　average weekday traffic (AWDT), 7

Base course, 83–84
　aggregates, 57
　cross section, 78, 81, 141
　inspection, 248
　material, 19
　　recycled, 229
　moisture penetration, 275
　reconstructing, 278
　shoulder, 81
　unit of measure, 210
　value-architecture, 55
Baseline, 36–37, 119, 180
Benefit-to-cost ratio, 20, 170
Bicycle, 9, 262–263
　bicyclist, 119, 159, 167, 173, 262–263
　facility, 197, 258, 262–263
　movements, 173, 206, 263
Bidding procedure, 217
Bituminous materials, 48, 60–62, 76–83, 88, 210, 251, 274, 278.
　See also Asphalt
BLM, see Bureau of Land Management
Boring, 50–51, 54
Box culvert, 142, 145, 274
Box girders, 146
Break-and-seat, 86–88, 228, 248
Bridge, 144–152. See also Structures
　adequacy, 7
　aesthetics, 97, 144, 177
　approaches, 127, 151, 225–226
　construction, 229, 240
　flyover, 169
　footing, 245
　geometry, 119, 144, 155
　glossary, 156–157
　ice formation on deck, 264
　inspection, 281–283
　large, 147–149
　layout, 251
　medium, 146–147
　metal, 62, 146–147
　orthotropic, 147, 151
　over waterway, 131, 155–156
　repairing/replacing/restoring, 150, 152, 234, 272, 274
　small, 145–146
　specialized, 149–150
　temporary, 149
　value-architecture, 19
Broken-back curve, 106–107
Budgeting, see Financing; Programming
Bureau of Indian Affairs, 36
Bureau of Land Management (BLM), 36, 67
Bureau of Reclamation, 36
Bureau of the Census, 38

Cable-stayed bridge, 149
CADD, see Computer-aided drafting

Index    295

and design
Capacity, 114
  congestion, 9
  constraints, 97, 159-160
  enhancing, 167-168, 264-265
  Highway Capacity Manual (HCM), 114, 160-161
  level of service (LOS), 114, 168, 187, 235
  planning for, 6
  requirements, 92, 195
Catch basin, see Drainage
Checking, 119-120, 142, 157
Classification of pavements, 76
Clearing the worksite, 125, 210, 241
Clear zone, 115-116
Coastal zone management program (CZMP), 36-37, 44
Collector road, 16
Community intrastructure, 5-6
Commuter parking lot, 10, 16, 26, 195-197
Compactors, 73, 88, 242-246, 251, 275-277
Computer-aided drafting and design (CADD), 14, 19, 39, 104
Construction materials, 48-63. See also specific items
Contour lines, 39-40, 44, 139
Contracts, 217-219, 226
Control line:
  horizontal, 37, 107, 109-110
  locating, 241
  offset, 156
  shown on plans, 119, 155
  vertical, 37, 108, 111, 116-117
Corridor, 6-7, 15-17, 20-21, 24, 27
Cost-effectiveness, 20-21. See also Value-architecture
  alternatives, 18
  commuter parking lot, 197

drainage, 132, 137
interchange, 170
materials usage, 78
pavements, 276
planning, 7
structures, 150-151
vegetation, 99
Council on Environmental Quality (CEQ), 95-96
Crack-and-seat, 86-88, 228, 248, 278
Critical path method, 221
Cross section, 114-116, 119
  highway, 56, 111-112, 114-116, 128, 138, 142, 155-156
  pavement, 78-81, 85
  railroad, 155
  waterway, 65, 70-71, 136, 142, 155
Culverts, see Drainage
Curb, 115-116, 153, 210, 263
Cutback, see Asphalt

Data envelopment, 12
Deeds, 201-202
Degree of horizontal curve, 107-110
Design hourly volume (DHV), 8
Design speed, 105-111, 155
DHV, see Design hourly volume
Digitized topographic model, 43
Directional split, 7-8
Drainage, 99, 114, 131-137, 176, 204, 210, 226. See also Water
  area, 64, 67, 132
  catch basins, 135-136, 210, 226, 238, 281
  grates, 238, 263
  condition, 47, 117, 161
  culverts, 64-65, 77-78, 94, 129-132, 134-136, 226, 245-246, 260, 279, 281
  box, 142, 145, 274

Drainage *(Continued)*
  design, 46, 66-67, 73, 113, 116-119, 128, 131-132, 134, 136-137, 141-142, 186, 274
  ditches, 64-65, 78, 110, 134-136, 141-142, 279, 281
  hydraulics, 67, 69, 71, 112, 132-136, 142, 144, 151, 226, 238, 283
  hydrology, 66-67, 71, 112, 132-134, 226, 241
  inspection, 246-247
  irrigation, 64, 67, 142, 186
  life cycle, 20
  maintenance, 272, 278-281
  mapping of, 41, 142
  pipe arches, 62, 145, 274
  pipes, 56, 62, 126, 134-136, 142, 156, 210, 226, 245-246, 260, 281
  storm sewers, 131, 135
  surface, 66, 110
    topsoil, 242
  underdrain, 78, 134, 154
  value-architecture, 19

Earthwork, 19, 41, 99, 110-112, 125-130, 139, 171, 210. *See also* Soils
  borrow, 127-129, 210
  embankment (fill), 19, 56, 65, 70, 78, 81, 111, 125-129, 131, 133, 139-140
    bridge approaches, 151, 155, 225-226
    compaction, 126-128, 139, 151, 223-226, 242-246
    instability, 56, 78, 152
    material, 41, 69-70, 94, 110, 112, 127-129, 242
    value-architecture, 55

  estimating, 46, 128-129
  excavation (cut), 56, 111-112, 125-129, 138-140, 210, 223-226, 240
    inspection, 242-243
    types, 126, 210
Easement, 202
Ecology, 65-70, 99
Emulsion, *see* Asphalt
Energy, 7, 259, 267
Environment, 91-102, 272. *See also* Aesthetics; Ecology; Noise; Recycling
  archeology, 125
  concern, 7, 11, 39, 49, 61, 71, 89, 126, 132, 136, 146, 241
  documents, 15, 93, 101
    environmental assessment, 96
    environmental impact statement, 95-96
  regulations, 95-96, 137, 218, 272, 283-284
  roadside, 176-177, 182
  water, 66-69
Enzymes, 77, 280
EPA, *see* U.S. Environmental Protection Agency
Erosion control, 65-66, 71-73, 94, 113, 137-143, 151, 176, 182, 226, 242, 244-247
Estimates, 112, 128-129, 139, 142, 157, 204, 209-211, 217-219, 241, 259
Exodermic concept, 147

Federal Emergency Management Agency (FEMA), 36, 67
Federal Highway Administration (FHWA):
  conversion to metric, xvi
  environmental guidance, 96, 99

wildlife, 180-184
research center, xv
right-of-way purchases, 200
structure advisories, 147, 152
traffic control study, 167
Fencing, 180-185, 210, 222, 285
Fill, *see* Earthwork, embankment
Financing:
   aesthetics, 92
   budget, 12, 115, 195, 272-274, 286
   economy of earthwork, 110
   interchanges, 170
   joint development, 29
   private, 10-11, 17
   rest areas, 190
   societal goals, 11-12
   value-architecture, 17-18
Fish and Wildlife Service, 67, 68
Flexible pavements, *see* Pavements
Floodplain, *see* Water
Flyover, 169
Freeze-thaw cycle, 84, 127
Freight and Fleet Operations, 290-291
Frost heave, 84

Gabions, 71-72, 153
Geographic information system (GIS), 38, 41, 67
Geologic hazards, 36-39, 119
Geometry, 103-121. *See also* Alignment; Control line
   aesthetics, 98
   coordination with:
      earthwork, 125, 139
      frost-heave prevention, 84
      geotechnical findings, 56
      structure design, 144, 155-156
      waterways, 64
   final, 113-119, 128
   improvement to, 7, 159, 161-168
   pavement, 78

   preliminary, 103-113
   safety aspects, 205
   traffic flow, 257
Geotechnical methods, 49, 54-57, 130
Geo-textiles (geosynthetics), 48, 62
   embankment, 152-153
   erosion control, 137, 141
   frost-heave mitigation, 84
   pavements, 82, 228
   slopes, 153, 247
   specifying, 208, 247
   subgrade, 83
   ultraviolet (UV) light, 247
   value-architecture, 55
   waterway applications, 72-73
Global positioning system (GPS), 37-38
Gradient, *see* Alignment
Graveled surface, 7, 22, 76-78, 82, 279-280
Grubbing, 125, 210, 222, 241
Guardrails (guide rails), 197, 211, 237-238, 260, 285

Hazards, *see* Safety
High-occupancy vehicle (HOV) lane, 9, 16
Highway Capacity Manual (HCM), 114, 160-161
Human factors, 167, 260-262
Hydraulics, *see* Drainage
Hydrology, *see* Drainage

Incident management, 7, 265, 290
Infrastructure, 6, 159
Innovation, 92, 149, 211-213, 258
Inspection, 227, 240-254, 283
Institute of Transportation Engineers (ITE), 209
Intelligent Vehicle/Highway Systems (IVHS), 289-292

Interchanges, 159, 170-174
  diamond, 171-174
  geometry, 112-114
  multiple use, 27
  ramps, 169
  retrofitting, 25
  right-of-way, 23, 38, 195
  traffic flow, 257
  type, 170-174
  value-architecture, 20
Interdiscipline, 101, 224
Intermodal Surface Transportation Efficiency Act of 1991 (ISTEA), 5, 12
Intermodal transfer, 7, 192-195
Intersections, 116-118, 159-171
  capacity, 114, 159-161
  conflicts within, 159-160, 259, 263
  converting to interchanges, 114, 257
  drainage, 117, 135
  geometry, 161-170, 206
  modifying, 114, 167-170
  profiles, 111, 116-118
IVHS, see Intelligent Vehicle/Highway Systems

Joint development, 23, 28-31, 201

$k$ factor, 7

Laboratory testing, 51-53, 60, 86-87, 249-251
LANDSAT, 42. See also Global positioning system
Level of service (LOS), 114, 168, 187, 235
Life cycles, 20, 170, 212
Lighting, 259, 285
Local roads, 7, 9, 76-78, 82, 105
Location:
  phase, 35, 103, 218
  guide to soils, 58-59
  studies, 18, 38-41, 44-47
  water impacts, 64-67, 73

Macadam pavement, 76
Maintenance, 15-16, 218, 271-288
  drainage, 131, 136, 280-281
  facilities, 25, 29
  interchanges, 170
  management, 272, 286
  pavements, 79-82, 110, 272-273, 274-280
  priorities, 272-274
  responsibility, 30, 168, 267
  roadside, 186, 258, 283-285
    rest areas, 191, 284
    vegetation, 179-180, 272, 281, 283-284
  roadway geometry impacts, 103, 112
  snow removal, 264, 280
  structures, 281-283
  underground water, 66
Manual on Uniform Traffic Control Devices (MUTCD), 167, 206, 234, 236-237, 258
Mapping, 35-47, 119, 142, 202, 204
Materials, see Construction materials
Metals, 62, 210
Metric units, xvi
Modular pavements, 82
Mosaics, 39, 41
Multidiscipline, 101
Multiple use of right-of-way, 23-31, 188, 202
MUTCD, see Manual on Uniform Traffic Control Devices

National Association of County Engineers (NACE), 209
National Bridge Inventory (NBI), 145, 150

National digital cartographic database, 38
National Environmental Policy Act (NEPA), 95
National Highway Institute (NHI), 118
National Oceanic and Atmospheric Administration, 37
National Park Service, 36, 67
NAVSTAR, see Global positioning system
NHI, see National Highway Institute
Noise, 7, 8, 13, 14, 81, 93, 100, 168, 178, 187-188, 233
Nonmotorized traffic, 262-263. See also Bicycle; Pedestrians

Orthophotographs, 39, 41
Orthotropic bridge design, 147, 151
Overloads, 79-80, 267

Park-and-ride, see Commuter parking lots
Pavements, 75-90. See also Shoulders
   aesthetics, 92, 177
   bridge approaches, 225
   classifications, 76
   condition, 7
   construction, 226-229
   cross section, 19, 78-85
   design, 85, 113
   ecological impact, 99
   flexible, 22, 76-86, 88, 228-229, 244, 273-278
      rutting, 79-80, 278
   frost action under, 84-85
   innovation, 212
   inspection, 247-251
   life-cycle, 20
   maintenance, 272-279
   markings, 79, 211, 232, 235, 258, 279
   materials, 49
   noise, 187
   payment items, 210
   recycling, 19, 83, 88-89, 94, 229, 249, 277
   rehabilitation of, 86-89. See also Break-and-seat; Crack-and-seat; Rubblizing
   removal of, 167-169
   rest stops, 189
   rigid, 76, 80-86, 88, 227-228, 274, 278-279
      pre-stressed, 81
      roller compacted, 81, 228, 249
   runoff, 135
   support, 49, 52, 56, 83, 127, 244
   utilities underneath, 184-185, 242
   wet, 263-264
Peak hour volume, 8
Pedestrians, 27, 119, 150, 159-161, 164-168, 173, 190, 196, 205-206, 235, 259-263, 279-280
Permafrost, 84-85
Photogrammetry, 41
Pipe arches, see Drainage
Pipes, see Drainage; Utilities, sewers
Planning process, 6-7, 15, 218, 282
Plans, 135-136, 155, 157, 180, 185, 202, 203-213, 217-219, 226, 228, 232, 235, 241, 246, 252
Polymers, 80
Portland cement, 48, 62, 210
   Association, 82
   concrete (PCC), 57, 72, 76, 80-87, 99, 210, 227-228, 248-249, 252, 273-274, 278
Positive guidance, 262
Private sector, 10-11
Proctor method, 52
Profile, see Alignment, vertical
Programming, 11-12, 190, 218

Railroad crossings, 159, 186, 259, 267
Reconnaissance, 35, 46-47
Recycling:
  flyash, 62, 89, 94
  pavements, 19, 83, 88-89, 229, 249, 277
    curbs and sidewalks, 153
  roadbed, 55
  rubber products, 94, 140, 152
  topsoil, 55, 222
  waste materials, 93-94, 222-223
Refraction seismology, 54
Rehabilitation, 86-89
Resistivity, 54
Rest areas, 26-27, 188-191, 274, 284-285
Retaining walls, 94, 97, 127, 144, 152-154, 226
Right-of-way:
  access, 7, 113, 118-119, 138, 186, 200-201
  acquisition, 19, 38, 46, 48, 112-113, 140, 152, 167, 170-171, 176, 200-201
  clearing, 181
  control, 113, 119, 186, 200
  deeds, 119, 201-202
  fence, 185, 283, 285
  joint development, 23, 28-31, 201
  limits, 223
  multiple use, 23-31, 99-100, 118, 131, 201
  restricted, 127, 169
  tract map, 201
  wildlife encroachments, 182-183
Rigid frame bridges, 145-146
Rigid pavements, 76, 80-86, 88, 227-228, 249, 274, 278-279
Riprap, 72-73, 136, 140, 210
Risk analysis, 136, 220
Roadside, 176-199. *See also* Erosion control; Joint development; Multiple use of right-of-way; Slopes; Vegetation
  aesthetics, 92, 97
  access, 186
  construction, 229-230, 252
  furniture (hardware), 112, 186, 197, 205, 238, 260, 263
  items, 113, 210, 252
  maintenance, 179-180, 272, 283-285
  noise abatement, 187-188
  plantings, 99, 176-180, 223, 272, 283-284
  rest areas, 188-191, 284-285
  wildlife control, 180-184
Robotics, 224-225
Rollers, *see* Compactors
Rubblizing, 86-88, 228, 248
Runaway truck ramps, 189
Running speed, 117

Safety, 205-206, 232-239
  accidents, 7, 110-113, 159-160, 181, 260, 291
    pedestrians, 260
  advanced driver information, 291
  bicyclists, 262-263
  constraints imposed by, 92, 97
  economics, 170
  enhancing, 114, 116, 272
  fencing, 182
  hazards, 98, 104, 110-115, 173, 180, 184, 186, 205-206, 224, 235-238, 260-264, 281
  intersection, 117, 161
  items, 211
  multiple use issues, 27
  roadside, 186, 238
    hardware, 205, 260

vegetation, 140, 205, 284
roadway cross section, 115
signing, 260
utility poles, 186
work zone, 232-237, 241
Sampling, 49-51, 130, 220, 223
Scheduling, 219, 221-222, 273, 286
SCS, *see* Soil Conservation Service
Seismic method, *see* Refraction seismology
Sewers, *see* Drainage; Utilities
Shoulders, 76-81, 92, 94, 115-116, 185, 234, 262, 264, 278-279, 283
SHRP, *see* Strategic Highway Research Program
Sidewalks, 116, 185, 210, 278-279
Sight distance, 77, 105, 108, 111, 113, 155, 161, 164, 177, 181, 186, 222, 234, 267, 283
Signs, 7, 77, 112-114, 119, 160-161, 167, 197, 211, 232-234, 240, 257-266, 272, 285-286, 290
Single point urban interchange, 171-173, 212
Site preparation, 222-223, 241-242
Slab bridges, 145-146, 156
Slopes, 78, 116, 126-129, 131, 134-141, 152, 177, 180, 186, 237, 244, 281. *See also* Earthwork; Erosion control
Socio-economic issues, 3, 11-12
Soil Conservation Service (SCS), 50
Soils, 48-59. *See also* Aggregates; Earthwork; Erosion control
classifications, 56, 58-59
compaction, 223-226, 242-246
construction problems, 136
density, 51, 127, 245
moisture content, 52, 137-138
report, 56-57, 112, 144, 156
sampling, 49-51, 130, 223, 245

slopes, 116
subgrade, 75, 83-84
testing, 49-53, 130, 227, 244-245
vegetation, 176
Specifications, 207-209
interpretation during construction, 204, 217, 219, 226, 228, 241-242, 246, 249, 252, 273
nonstandard, 142, 157, 211
special provisions, 140, 157, 208-209, 211, 218, 246
standard, 128, 157, 203, 207-208, 224, 259
range, 227
supplemental, 218
Spiral curves, 107, 113
Staffing, 240-241
State plane coordinate system, 37, 45
Stationing, 107, 119, 128, 142, 155-156
Strategic Highway Research Program (SHRP), 75, 79
Stream relocation, *see* Water
Structures, 144-158. *See also* Bridges; Retaining walls
construction, 139
inspection, 251-252
excavation for, 126, 243
foundation, 51, 56, 243-244
root piles, 154
geometry, 113, 128
items, 210, 229
life cycle, 20
maintenance, 281-283
materials, 49
noise conductance, 188
plans, 155, 157, 204
rest area, 189
tunnels, 150
underneath, 26
value-architecture, 19

Structures *(Continued)*
  waterway crossing, 65, 131, 155–156
Sub-base, 19, 56, 78, 81, 83, 154, 210, 248
Subgrade, 52, 55–56, 75–85, 110–111, 128, 141, 185–186, 210, 226–227, 244, 248, 275, 278
Subsurface investigation, 48–50
Superelevation, 106–107
Surveying, 35–38, 114, 119, 201, 203, 222, 241, 251

Tangents, *see* Alignment, horizontal
Testing, 49–53, 57, 60, 83, 86–87, 130, 220, 227–228, 241, 244, 249–252
TIGER file, 38
Topography (terrain), 35–36, 44, 46, 67, 104–105, 110–119, 126, 139, 177, 180, 184
Tract map, 201
Traffic, 257–270. *See also* Capacity
  control plan (TCP), 235
  geometric influence, 110–113, 115
  incident management, 265–267, 289–291
  intersection conflicts, 159–171, 205–206
  maintaining and protecting, 146, 204, 235–237, 277
  monitoring, 25, 290
  multiple use issues, 27
  noise, 168, 187, 233
  pavement:
    adequacy, 78, 81–82
    interaction, 77, 79–80
  volume, 7–8, 99, 116, 145, 181, 189, 191
  weather effects, 263–264
  wildlife conflicts, 181–182
  worksite, 233–234, 277
Transportation Research Board (TRB), xv, 75, 99
Transportation systems management (TSM), 16
Truck escape ramps, 237–238
Truss bridges, 146–147, 156
Tunnels, 150

U.S. Air Force, 276
U.S. Army Corps of Engineers, 20, 36, 89, 276
U.S. Coast Guard, 67, 69
U.S. Department of Agriculture, 139. *See also* Soil Conservation Service; U.S. Forest Service
U.S. Department of Commerce, 37. *See also* National Oceanic and Atmospheric Administration
U.S. Department of the Interior, *see* Bureau of Indian Affairs; Bureau of Land Management; Bureau of Reclamation; Fish and Wildlife Service; National Park Service; U.S. Geological Survey
U.S. Department of Labor, 210
U.S. Environmental Protection Agency (EPA), 67, 68, 95, 96, 136
U.S. Forest Service (USFS), 23, 36, 67
U.S. Geological Survey (USGS), 36, 38, 44, 50
Utilities:
  above ground, 27, 39, 46, 125, 160, 185, 237
  coordination, 27–28, 118, 180, 184, 186, 210, 230, 277
  reconnoitering for, 46–47
  relocating, 184, 186, 252
  sewers, 46–47, 184

underground, 39, 46–47, 125–126, 136, 160, 184–185, 219

Value-architecture, 17–20, 55, 85, 110, 205–206
Value engineering, 17–18, 20, 209
Vegetation:
  aesthetics, 140, 176
  ditches, 134
  erosion control, 140–141, 176
  establishment of, 138, 281
  fires and burning, 178, 241–242
  measurement units, 210
  removal of, 180–181, 186, 222, 279, 284
  riparian, 69, 71–72
  roadside, 99, 176–180, 182, 187, 272, 283–284
  trees, 92, 140, 177, 186–187, 205, 279, 284
  weeds, 242
Vertical curves, *see* Alignment, vertical

Water, 64–74
  availability, 184, 226
  floodplains, 36, 67–70, 93, 95, 112, 131, 136
  irrigation, 64, 67, 142, 281
  quality, 218
  stream relocation, 65–73, 94, 134, 136, 142, 152
  underground, 129, 132–133
  waterways, 36, 39, 44–45, 64–74, 91, 96, 110–116, 131, 144–152, 155, 184, 245
  scour, 71–73, 283
  wetlands, 36, 39, 44–45, 64–69, 94–95, 137
Wildlife, 24, 69–70, 95–96, 99, 134, 180–184, 222